AEROENGINE FUNDAMENTALS AND LANDSCAPE IN INDIA

A Way Forward

AEROENGINE FUNDAMENTALS AND LANDSCAPE IN INDIA

A Way Forward

Air Vice Marshal **Suresh Singh** AVSM VSM (Retd)

Introduction by
Air Marshal **Anil Chopra** PVSM AVSM VM VSM (Retd)

Centre for Air Power Studies
New Delhi

in association with

KW Publishers Pvt Ltd
New Delhi

CENTRE FOR AIR POWER STUDIES

VISION

To be an independent centre of excellence on national security contributing informed and considered research and analyses on relevant issues.

MISSION

To encourage independent and informed research and analyses on issues of relevance to national security and to create a pool of domain experts to provide considered inputs to decision-makers. Also, to foster informed public debate and opinion on relevant issues and to engage with other think-tanks and stakeholders within India and abroad to provide an Indian perspective.

Contents

Preface

The book emphasises the strategic importance of indigenous design and development of a 110-130 KN turbofan aeroengine for use on military as well as commercial aircraft. It covers the fundamentals of aircraft engines in the simplest possible way without using any complex mathematical expressions. The book is divided into 10 chapters, covering all the important aspects of modern aeroengines. It can be useful for engineering students, commercial and military pilots, and practising engineers, besides Indian industry leaders and decision-makers. Chapter I covers the global history of aeroengine design and development for commercial and military aircraft. Chapter II explains the fundamentals of aeroengines in details that can be assimilated by anyone. Chapter III encompasses the historical perspective of aeroengine design and development in India and the circumstances under which the first Asian combat aircraft, the Marut, manufactured by HAL, had to be prematurely retired in the absence of a suitable aeroengine. Chapter IV covers the various aspects of aeroengine maintenance, repair, and overhaul in India and the various processes involved. Chapter V covers the current status of aeroengine development, including industrial gas turbine production in India, with a view to showcasing the potential of engine manufacturing capabilities in the country. Chapter VI briefly covers the Gas Turbine Research Establishment (GTRE), its efforts towards the Kaveri engine development in India, and the difficulties experienced. Chapter VII covers the incorporation of advanced technologies in modern aeroengines across the world pertaining to the reduction of pollutant gases, noise, and fuel consumption. Chapter VIII covers the various parameters monitored during the ground testing and the air flight testing of aeroengines and the requirement of a suitable air platform for air flight tests. Chapter IX covers the potential of the aeroengine business in India for commercial and military aeroengines in detail. Chapter X discusses the various options to ensure a time-bound programme for the design and development of aeroengines in India. The most feasible option for the development of the indigenous aeroengine is under a public-private partnership under the overall supervision of the PMO.

Introduction

India's great success in space launch vehicles has often invoked questions about its inability to successfully make an aeroengine. Many people, including some intellectuals, do not fully understand the complexities of aeroengine operation. These engines have to operate efficiently from zero platform speed to high Mach numbers; from zero altitude to higher atmosphere; rotate at very high Revolutions Per Minute (RPMs) and temperatures; withstand high positive and negative 'g' manoeuvres; and do all this repeatedly for the entire life of the aeroengine, which could be a few thousand hours or mission cycles. While the manufacturing complexities may be similar, the aeroengine designs for fighter aircraft and large transports and airliners are different.

There is just a handful of successful aeroengine manufacturers around the world. The top four global aircraft turbofan engine manufacturers are Pratt & Whitney (P&W), Rolls-Royce, General Electric (GE) Aviation and Safran. GE and Safran of France also have a joint venture called CFM International. P&W also has a joint venture, International Aeroengines with the Japanese Aeroengine Corporation and MTU Aeroengines of Germany. Pratt & Whitney and General Electric have a joint venture, Engine Alliance, selling a range of engines for aircraft such as the Airbus 380. Honeywell Aerospace is another aeroengine manufacturer of, among others, smaller engines, including for cruise missiles.

The US-based Williams International, Russia's Aviadvigatel and Ukraine's Ivchenko-Progress are other manufacturers. PowerJet is a joint venture between Snecma (Safran) and NPO Saturn of Russia.

Chinese Aircraft Corporations at Shenyang, Xi'an, and Guizhou also make aeroengines. Most engine manufacturers make engines for both civil and military aircraft. Russian commercial aircraft often use Western aeroengines. The Chinese depend on Russia for fighter jet engines and Western companies for airliner engines. Despite years of investment in Research and Development (R&D), China continues to struggle in its engine development. Several major manufacturers have formed joint ventures to access both technology and markets.

India's GTRE GTX-35VS Kaveri turbofan engine has been under development by the Gas Turbine Research Establishment (GTRE), a lab under the Defence Research and Development Organisation (DRDO). The Kaveri was originally intended to power production models of the HAL Tejas Light Combat Aircraft (LCA) developed by Hindustan Aeronautics Limited (HAL). However, the Kaveri programme failed to satisfy the necessary technical requirements or keep up with its envisaged time-lines and was officially delinked from the Tejas programme in September 2008.

After being on hold for nearly a decade, a dry variant of the Kaveri engine is now being developed to power the DRDO Ghatak by 2026. Meanwhile, GTRE kept working on the engine on its own. In June 2022, it was reported that the combustion in the redesigned Kaveri engine was now much more stable. The Indian firm Godrej & Boyce has been given the contract to manufacture six more engines for further tests. In February 2023, high-altitude tests of the engine were reportedly conducted successfully in Russia.

Despite long time-lines and delays, Indian scientists and engineers have had great developmental learning from the Kaveri and have collected a lot of raw data that can be put to use. Over the years, India has produced and overhauled many aeroengines of Russian and Western origin under licence. The manufacturing and testing capabilities are well in place. Many components have been sourced locally, albeit India was dependent for many critical metals and materials on the Original Equipment Manufacturers (OEMs). India will soon start manufacturing the GE-414 in India. But, effectively, it may still be some form of licensed production, though it will perhaps bring new production techniques and technologies and build a sub-systems eco-system.

If India has to become a significant global aerospace player, it must develop its own aeroengine with full intellectual property rights. Preferably, India should invest much more in research and development of a new engine. If required, India could hire highly paid international specialists. But in view of the lack of technologies, a safer alternative may be to design and evolve the aeroengine with the help of a major aeroengine manufacturer. Even this route will cost a lot of money. The decision must be taken quickly.

Since the subject of aeroengine development is of immense importance, we, in the Centre for Air Power Studies (CAPS), decided to get a senior aeroengine specialist to carry out a study to evolve a good roadmap for engine development. India's Tata Group was very kind to support such a study by instituting a Research Chair of Excellence, and funding the study.

Air Vice Marshal Suresh Singh, who has had an extensive experience on Maintenance, Repair, and Overhaul (MRO) of British, French and Russian origin aeroengines on the Indian Air Force's (IAF's) fighter, transport, and helicopter fleets, was chosen to do the study. He was supported with interactions with many other aeroengines specialists, and given the opportunity to visit research and industry establishments. Each chapter was peer reviewed.

The study report covers in detail the fundamentals of aeroengines, the historical aspects of aeroengine development, and the current MRO in India. The author then went into the current status of India's aeroengine development, and the work done by GTRE. He assessed the potential of the aeroengine business in India and the importance of incorporation of new advanced technologies. He looked into the ground testing and flight testing route. And, finally, he has proposed the way forward for India.

The Indian government is driving *atmanirbharta* in defence in all earnest, at the highest levels. The aeroengine is now one of the most important systems in the cross-hairs for development. India has been talking with major aeroengine manufacturers. There are issues of total thrust, thrust-by-weight, level of technology, and costs. All these will lay the foundation to power next-generation fighter jets like the LCA Mk2 and the Advanced Medium Combat Aircraft (AMCA). The data acquired from the Kaveri development may also be gainfully used. The process will also support understanding and developing some core technologies and materials, including superalloys. The private sector would have significant involvement. A major offshoot would be to later develop a commercial aircraft engine.

Aeroengine Fundamentals and Landscape in India: A Way Forward, is a very well researched report. The author has years of experience on aeroengines. He has been both a maintainer and an operator. The research clearly brings out the current status of India's developmental capabilities and elucidates the best options forward. The study will be very useful for planners at the Ministry of Defence, and Air Headquarters (HQ), as well as for the scientific community. It will also make a great reference document for aerospace practitioners and researchers.

Air Marshal **Anil Chopra**
Director General
Centre for Air Power Studies

Acknowledgements

I wish to express my gratitude to Air Marshal Anil Chopra PVSM AVSM, VM, VSM (Retd), Director General, Centre for Air Power Studies (CAPS), Air Vice Marshal Anil Golani (Retd) Additional Director General, CAPS, and Air Vice Marshal S C Luthra AVSM VSM (Retd) for entrusting me with the responsibility to undertake research on the very relevant and important subject "Aeroengine Landscape in India". I sincerely thank Air Marshal P Kanakaraj, PVSM AVSM VSM (Retd), for his valuable suggestions, encouragement, and necessary guidance during my research work. I also sincerely thank Professor (Dr) A P Haran, Air Vice Marshal Sarad K Jain (Retd), Air Vice Marshal Vikash Dwivedi, Dy SMSO, HQ MC, Air Vice Marshal Anoop Ghosh (Retd), Group Captain Pradeep K Sahoo (Retd), Group Captain K Giri and Squadron Leader Anjali Joshi (Retd) for their professional review and valuable suggestions for the completion of my research work. I also thank Group Captain (Dr) Swaim Prakash Singh, Senior Fellow, CAPS and his staff for providing the necessary help and support in editing and publishing the book. I express my gratitude to my wife, Kumud K. Singh, and my daughters, Payal Singh and Sanskriti Singh, for continuously encouraging me to undertake this research work wholeheartedly.

Abbreviations

AC	Alternating Current
ADA	Aeronautical Development Agency
ADVENT	Adaptive Versatile Engine Technology
AETD	Adaptive Engine Technology Demonstrator
AETP	Adaptive Engine Transition Programme
AIESL	Air India Engineering Services Limited
AMCA	Advanced Medium Combat Aircraft
ASR	Air Staff Requirement
ASRS	Aeroengine Steering Revolution System
BEL	Bharat Electronics Limited
BH	Baker Hughes
BHEL	Bharat Heavy Electrical Limited
BPR	Bypass Ratio
CAMC	Composite Carbon Matrix Composites
CEMILAC	Centre for Military Airworthiness and Certification
CFRP	Carbon Fibre Reinforced Plastic Composite
CIAM	Central Institute of Aviation Motors
CMC	Ceramic Matrix Composite
CSIR	Council of Scientific and Industrial Research
CVD	Chemical Vapour Deposition
DC	Direct Current
DPSU	Defence Public Sector Undertaking
EBW	Electron Beam Welding

EDM	Electric Discharge Machines
FAA	Federal Aviation Administration
FADEC	Full Authority Digital (Electronic) Engine Control
FDI	Foreign Direct Investment
FSED	Full-Scale Engineering Development
GE	General Electric
GETL	General Electric Triveni Limited
GFRI	Gromov Flight Research Institute
GSLV	Geostationary Launch Vehicle
GTRC	Gas Turbine Research Centre
GTRE	Gas Turbine Research Establishment
HAL	Hindustan Aeronautics Limited
HP	High Pressure
HYTEC	Hybrid Thermally Efficient Core
IAF	Indian Air Force
IGCAR	Indira Gandhi Centre for Atomic Research
IICT	Indian Institute of Chemical Technology
IPR	Intellectual Property Rights
ISA	International Standard Atmosphere
ISRO	Indian Space Research Organisation
JNARDDC	Jawaharlal Nehru Aluminium Research Development and Design Centre
KN	Kilonewton
LCA	Light Combat Aircraft
LP	Low Pressure
LRU	Line-Replaceable Units
MMR	Multi-Mode Radar
MRO	Maintenance, Repair, and Overhaul
NCLT	National Company Law Tribunal
NGV	Nozzle Guide Vanes
OEM	Original Equipment Manufacturer
P&W	Pratt & Whitney
PAC	Public Accounts Committee

PMO	Prime Minister's Office
PSLV	Polar Satellite Launch Vehicle
PVD	Physical Vapour Deposition
R&D	Research and Development
RPM	Revolutions Per Minute
RWR	Radar Warning Receiver
SAF	Sustainable Aviation Fuel
SFC	Specific Fuel Consumption
SHP	Shaft Horse Power
SPJ	Self-Protection Jammer
TBC	Thermal Barrier Coating
TIG	Tungsten Inert Gas
ToT	Transfer of Technology
TRL	Technology Readiness Level
USAF	United States Air Force
YSZ	Yttria Stabilised Zirconia

1

Prolegomenon

The invention of the jet engine was a monumental step in the history of air transportation. This also ushered in a new era in the arena of military aviation. Jet engines made air transportation the most reliable, safest, and fastest mode of travel in the area of domestic and international transportation segments. For the military, it has provided unprecedented speed for decisive action in the eventuality of war while also assisting civil society during natural calamities. The evolution and improvement in jet engine design in terms of size, weight, power, fuel efficiency, reduction of air pollutants, noise level, and reliability are continuous processes that are being undertaken by original equipment manufacturers (OEMs).

The first operational jet engine was designed by Hans von Ohain of Germany, though credit for the invention was given to Frank Whittle of Britain. Frank Whittle registered a patent for the turbojet engine in 1930 but did not perform a flight test until 1941. Ohain, while pursuing his doctorate at the University of Gottingen, formulated a theory of jet propulsion in 1933 and was granted a patent for his turbojet engine in 1936. Hans von Ohain worked for Dr. Ernst Heinkel at the Heinkel company in Rostock, Germany. Ohain built a factory-tested demonstration engine in 1937. The first jet aeroengine (He 178) designed by Dr Ernst Heinkel and Anselm Franz of Germany flew on a Me 262 fighter aircraft in Germany on August 27, 1939. Meanwhile, in England, Frank Whittle also invented a jet engine completely on his own, and this engine was installed on the Gloster Meteor. The aircraft was not used for combat due to its insufficient speed; however, it was used for homeland defence. The British shared Whittle's technology with the United States of America, allowing General Electric (GE) to build jet engines for America's first fighter aircraft, the Bell XP-59. The Whittle technology was also used by Rolls-Royce for developing newer aeroengines, such as the Nene, in 1944. Rolls-Royce sold Nene aeroengines to the Soviet Union. The Soviet version of Nene aeroengines powered the MiG-15. Pursuant to Germany's surrender in 1945, GE and Pratt & Whitney (P&W) got access to German technology and used it to improve the British-origin Nene aeroengine. In fact, P&W combined the two aeroengines (British and German) into one. The aeroengine included two compressors, each rotated independently by their own turbines, with the inner one giving high compression for better performance. Pratt and Whitney developed J-57 aeroengines for the commercial airlines Boeing 707 and Douglas DC. It was indeed a successful engine that also got inducted into the US Air Force in 1953. The basic design and configuration remain the same; however, in terms of engine efficiency, reliability, and use of advanced materials for various modules and sub-systems of the jet engine, they have gone through a phenomenal transformation. Aeroengine design and development involve a complex process and involve the application of advanced technologies in the fields of thermodynamics, material science, flow dynamics, and related technologies, and thus remain confined to very select industries in predominantly four countries, i.e., the USA, France, UK, and Russia only.

The Government of India (GoI), while taking a decision to design and develop a light combat aircraft way back in 1985, also sanctioned a project pertaining to the design and development of an aeroengine suitable for installation on the Light Combat Aircraft (LCA) Tejas. The responsibility for the design, development, and prototype manufacturing of an aeroengine was assigned to the Gas Turbine Research Establishment (GTRE), Bengaluru. The first aeroengine, named Kaveri, was designed and developed by GTRE; however, the engine could not produce the specified thrust as envisaged during the ground/air test of the aeroengine, and the weight of the aeroengine was also higher than the specified value.

The basic construction and functioning of jet engines for civil and military aircraft are similar. However, the size, weight, thrust, and system of integration on military combat aircraft and civilian commercial airliners differ primarily due to the characteristics of the role and requirements of military combat aircraft. Chief differentiating factors are faster acceleration and deceleration, higher manoeuvrability, shorter airstrip take-off and landing, operation on semi-prepared runways, augmented thrust through the afterburner, etc.

The development of military engines for fighter aircraft started with pure jet aeroengines; however, to improve fuel economy and reliability, the compressor design was improved with longer fan blades and wider chords. However, to prevent the bending of compressor blades and touching adjacent blades at high rotational speeds, the designers provided additional strengthening and support points to provide physical support on the blade length. While the British call them clappers, the French terminology specifies them as snubbers. A number of design alterations were attempted for higher atomisation of the fuel and an ideal air-to-fuel mixture to achieve better fuel efficiency. The evolution of the aeroengine is a continuous process, and it's likely to be more challenging in the future to achieve better fuel economy, less noise, a reduction of air pollution, less maintenance, lower inspection and downtime, and higher reliability.

The general trend in aeroengine design is to improve the air compression ratio and the thrust-to-weight ratio through the design of high bypass ratio engines. However, the high bypass ratio design is highly beneficial in terms of efficiency for the high thrust of the large commercial engine, whereas the military aeroengine for fighter aircraft uses a low bypass engine to keep the engine size smaller. This is mainly because the engines are installed inside the airframe for high agility, adequate stability, and higher manoeuvrability. However, commercial aeroengines are installed under the wing of the aircraft and thus can accommodate a larger engine frontal diameter with a higher bypass air passage.

The material being used for the aeroengine casing is predominantly aluminium magnesium alloy, primarily due to its low weight and high strength characteristics. The trend for the use of aluminium-magnesium alloy started during the years 1940-50 due to the abundant availability of magnesium alloy, which was earlier utilised to produce military hardware during the Second World War. Upon the end of the Second World War, there was a drastic reduction in demand for aluminium-magnesium alloy for military hardware.

In India, the majority of military aircraft and all private, civil, and commercial aircraft have been procured ex-abroad. The engine was provided by the OEMs as a standard fit, even though in most advanced countries, the OEMs of aircraft are required by rule to provide alternative engines for a specific aircraft to avert monopolistic business practices in the future. However, in the Indian military as well as the civil segment, the requirement of an alternate aeroengine was probably not insisted upon during the procurement phase, and hence we adhered to the standard-fit engine by the OEMs. The need to develop an aeroengine for military or commercial aircraft was not considered important in India primarily due to lower priority from the government, inadequate domain knowledge, a lack of risk appetite for research and development, the high cost involved in the research, design, and development of indigenous aeroengines, and the long lead time for testing, validation, certification from competent authorities, and approval of OEMs for installation on the aircraft. However, this aspect served the business interests of OEMs for their continuous revenue earnings aspects during the entire service life of the aeroengine; it also forced the Indian customers in the commercial sector and even the GOI for military aircraft operations into permanent dependence on OEMs.

As we are aware, there are only 4-5 OEMs in the world that have successfully designed, developed, and manufactured reliable aeroengines for civil as well as military aircraft. All the OEMs have invested a huge amount of resources in the design and development of various technologies, including the nurturing of ancillary industries and managing the manufacturing environment pertaining to aeroengines. They are reaping the benefits of these technologies through the sale of new aeroengines to various customers and also in the form of technology support to various customers for the entire life span of the aeroengine in terms of the supply of spares and undertaking maintenance, repair, and operations (MROs). They are unwilling to share the core technologies with any aspiring countries in view of the detrimental effect on their futuristic business interests. The interesting feature is that even though these OEMs have some joint ventures among themselves for new projects or new engine development, they still do not share their patented technologies with each other.

It is essential to note that none of the aeroengine design institutions and OEMs will share any of the core technologies at any cost with any new aspiring country or institutions to enter the exclusive club since they intend to retain their strategic position in design, manufacturing, and a continuous source of revenue earnings from after-sale services. The only option for the new entrants is to develop these technologies in-house in collaboration with institutions of national repute or to co-develop the technologies with the established OEMs on a risk and revenue-sharing basis.

GTRE has been actively involved in the design, development, and prototype manufacturing of military aeroengine for the last three decades. It has got fair technical know-how for 60-70 KN of aeroengine while working on the Kaveri engine. However, India needs to move forward towards design, development, prototype manufacturing and ground/air testing of 110-130 KN core aeroengine suitable for Advanced Medium Combat Aircraft (AMCA) for which GTRE is not well equipped in terms of technical know-how and in-house development may take a longer time frame.

The alternative option is to involve top Indian industrial enterprises in the private and public sectors and academic and research institutions for collaborative engagement in the design, development, and prototype manufacturing of low-bypass turbofan 110-130 KN thrust modular core aeroengines for dual use to ensure the accelerated pace of development. The project is to be headed by a person specially selected with adequate domain knowledge to coordinate the entire project and ensure adherence to a strict time schedule. The project is to be funded by the government, and thus, all the intellectual property rights generated for various line-replaceable units (LRUs), sub-assemblies, major assemblies, and technologies will belong to the government.

The work for the design, development, and prototype manufacturing of an aeroengine may be divided into several parts, and the complete responsibility for the development of specific parts or technologies should be assigned to a specific entity based on its competency, proven track record, knowledge base, aptitude for the research work, and risk appetite.

It is pertinent to note that a large range of technologies has been developed by the Indian Space Research Organisation (ISRO) in the arena of material science, machining, micrometry aerodynamics, automation, test benches, sealants, and polymers, which can be useful for the design and development of aeroengines as well. Moreover, the Indian industrial enterprises have also added advanced machining facilities, including micrometry and precision manufacturing capability, material processing, electron beam welding (EBW), broaching machines, robotic plasma spray machines, the most sophisticated digital balancing machines, electric discharge machines (EDM), high isostatic presses (HIP), physical vapour deposition (PVD) and chemical vapour deposition (CVD), heat treatment, and additive manufacturing capabilities. It is important to note that a large number of Indian companies are already supplying a number of critical spare parts, such as main bearing support housings of aeroengines, to foreign OEMs as per the engineering or manufacturing drawing (build to print) and the specified materials, e.g., the machining of a GE-404 engine high-pressure turbine disc being undertaken by a Chennai based private company. There are presently approximately 2000 companies in India supplying a large range of engineering-related products to foreign-based OEMs for aircraft and aeroengine parts and systems.

Based on extensive research in the civil and military aeroengine domains in India, it has emerged that the following government and private sector companies are engaged in the ab initio design, development, and prototype manufacturing of aeroengines.

(a) GTRE, Bengaluru: Research and development (R&D) for medium power low bypass jet engines (Kaveri).

(b) Hindustan Aeronautics Limited (HAL) (ERDC), Bengaluru: Single-stage centrifugal compressor engine of 100 kW (GTSU-110).

(c) Paninian India Private Limited, Bengaluru: Design of axial and centrifugal compressor aeroengine of smaller ratings mainly for auxiliary power units for commercial aircraft and other miscellaneous applications.

The following Defence Public Sector Undertakings (DPSUs) are engaged in under-license manufacturing or MRO of military engines in India.

(a) HAL, Bengaluru: Adour Mk871-07 (HAWK Mk-132), Mk811, Mk804E (Jaguar and Hawk), TM3332B2 (Dhruva helicopter), TM3332M2 (Cheetal helicopter), Dart5333-2 and 536-2T (Avro), Garret (Dornier)

(b) HAL, Koraput: R-25 (MiG-21), R-29 (MiG-23/27), RD-33 (MiG-29), AL-31 (Su-30)

(c) Verman Aviation Pvt Ltd., Bengaluru: PT-6 variants (Cessna, Beech-Bonanza and other small aircraft)

(d) Base Repair Depots of the Indian Air Force (IAF) are also engaged in maintenance, repair, overhaul and defect investigation of a large range of fighter, transport and helicopter aeroengines such as M53P2 (Mirage-2000), Viper 22-8 (Kiran trainer ac), R-29 (MiG-23/27), AI-20D (AN-32), TV-2, TV-3 MT, VK, VM and VK-2500 (medium lift helicopters) under arrangement with transfer of technology (ToT) from the respective OEMs from Safran, (France), Rolls-Royce, (UK), Motor Sich (Ukraine), and Klimov Plant (Russia).

The following Indian companies are engaged in MRO of civil and commercial aircraft aeroengines:

(a)   Air India Engineering Services, New Delhi and Mumbai: CFM-56-5B,7B (Boeing 737, A320 family, A340-200/300), GE CF6-80C2, GE-90-110 (Boeing 777), V2500A1 (Boeing 737, A320 family), PW4000-94 (A300-600, A310-300, Boeing 747-400, Boeing 767-200/300)

(b)   Verman Aviation Pvt Ltd, Bengaluru: PT-6 variants (Cessna, Beech-Bonanza and other small aircraft)

(c)   Safran India, Hyderabad: Yet to commence technical services.

(d)   GE, USA and HAL, Bengaluru: Yet to start services

There are approximately 41 GE-404-IN-20 aeroengines already procured for LCA Tejas Mk1, and a contract has been signed for another 99 GE-414 aeroengines for 83 LCA Tejas Mk II aircraft. The Aeronautical Development Agency (ADA) has already started the design and development process for the Advance Medium Combat Aircraft (AMCA). The IAF may procure up to 126 AMCA aircraft (seven squadrons). Thus, there will be 249 indigenous combat aircraft operating with the IAF in the next 15-20 years. As per general norms, approximately 2.5 to 3 aeroengine are required during the entire service life of the combat aircraft, and thus, there will be a requirement of $249 \times 2.5 = 622.5$, or 623 aeroengines, in respect of indigenous combat aircraft. Hence, there is likely an expenditure of US\$ 4.36 billion (US\$ 7 million/engine) for sustenance costs in respect of aeroengines for indigenous aircraft. Moreover, the life cycle support requirement in terms of spares and MRO can also be projected to the tune of approximately US\$ 2.8 billion (64 per cent of sustenance cost).

In the civil and commercial segments presently, there are more than 700 aircraft in India being operated by various operators, which is likely to increase to 1,700 aircraft by 2030. Indigo placed an order for 310 single-aisle aircraft from Airbus A320 neo, A321 neo, and 321 XLRs aircraft and 620 Leap-1A engines in 2021. Air India announced an order for 400

Airbus and Boeing 737 aircraft and 800 CFM-LEAP 1A engines at the recent Aero-India event in Bengaluru in 2023. Thus, there will be a huge demand for new aeroengines (approximately 1,440 aeroengines of the CFM-56, LEAP-1A/B class) and MRO business for these engines in India to the tune of US\$ 15.5 billion by 2030.

Thus, it can be safely estimated that in the near future, the combined volume of MRO business for the aeroengine in the military and civil commercial segments is likely to exceed US\$ 21.9 billion by 2030. The volume of business is further likely to increase in view of the GOI's push for the UDAAN project, which involves regional connectivity for the tire-II cities in the near future. In all likelihood, the establishment of adequate MRO facilities in India for the military and civil commercial engines will reduce the MRO cost substantially compared to MRO at OEM premises abroad, besides reducing lead time and transportation costs, providing employment opportunities to at least two million workers directly and indirectly, and saving huge amounts of foreign exchange from the na

2

Fundamentals of the Aeroengine

The aeroengine, in its simplest form, can be explained as a gas turbine engine that is operated by gas, which is a product of the combustion of fuel mixing with air and burning. It works on the basic principle of gas expansion at elevated temperatures at constant pressure, contrary to other internal combustion engines, like automobile engines, which work on the expansion of gases at an elevated temperature at a constant volume. Moreover, in the case of aeroengines, the four sub-cycles of operation, i.e., induction of air, heating up or temperature rise, expansion, and exhaust, take place simultaneously, unlike in other internal combustion engines, where the four sub-cycles of engine operation take place one at a time. The changes in temperature, pressure, and volume of air take place at various stages of the engine's operation, such as during compression. When the work is done to increase the pressure and decrease the volume of the air, the temperature will rise proportionally. In the combustion phase, the compressed air enters the combustion chamber, where the fuel is mixed with the air and burnt. The temperature of the air will rise along with the volume; however, the pressure remains constant. In the expansion phase, the work is extracted from the air stream by the turbine blades, wherein the temperature and pressure will decrease; however, the volume of the air will increase. The hot gas at high velocity will exit from the jet nozzle assembly and thus provide the thrust. The basic principle of gas flow through the convergent and divergent ducts is explained in Figs 2.1 and 2.2:

Fig 2.1: Convergent Duct

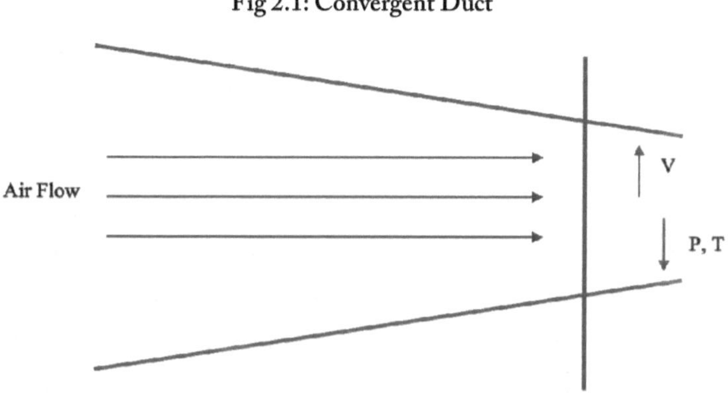

Fig 2.2: Divergent Duct

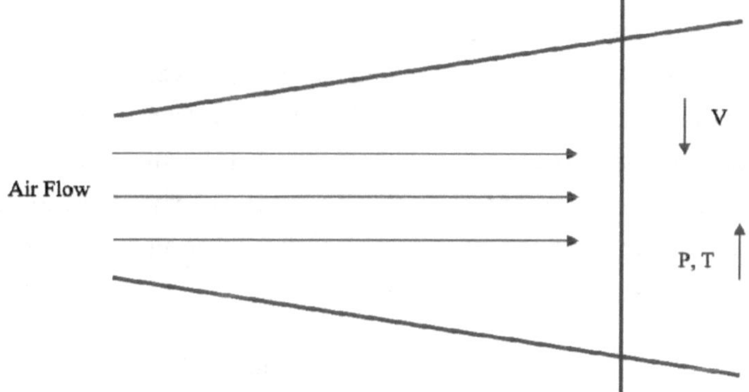

Convergent- Divergent Nozzle

It is evident from the above figures that in a convergent duct for the subsonic flow pattern, when the hot gases pass through the duct, the velocity increases, whereas the pressure and temperature drop; however, when the hot gases pass through a divergent duct, the velocity of the hot gases reduces, whereas the pressure and temperature increase. The entire process reverses when the gas flow velocity operates in a supersonic flow pattern. As evident, since the turbojet engine is a heat-based engine, the higher the temperature in the combustion chamber, the higher the expansion of the gases, resulting in higher thrust. However, since the material of the combustion chamber and the turbine blades have severe limitations, the gas temperature entering the turbine segment needs to be regulated in such a way that, under no circumstances, the hot gas temperature at the entrance of the turbine segment exceeds the design value. It is important to note that a definite change occurs in the temperature, pressure, and volume of the gas while it is flowing through the engine, as per the Boyle and Charles laws. The combination of the gas laws, $PV \propto T$, i.e., the multiplication of pressure and volume of the gas at different stages of the aeroengine, remains proportional to the absolute temperature of the air at the respective stage of the engine. The gas turbine's working cycle includes suction of the ambient air, compression through the compressor, combustion in the combustion chamber, and exhaust through the jet nozzle. The variation in velocity and pressure also takes place at various stages of operation. In the compression phase, the air pressure continuously rises from the lower to the higher stages of the compressor, but the velocity remains the same; however, in the combustion chamber, the temperature rise will take place, and the expansion of the air will take place, resulting in a higher velocity prior to entering the first stage of the turbine. In order to understand the general operation of the aeroengine, the following engine cut-view diagram can be useful.

Fig 2.3: Cut-View of a Functional Engine

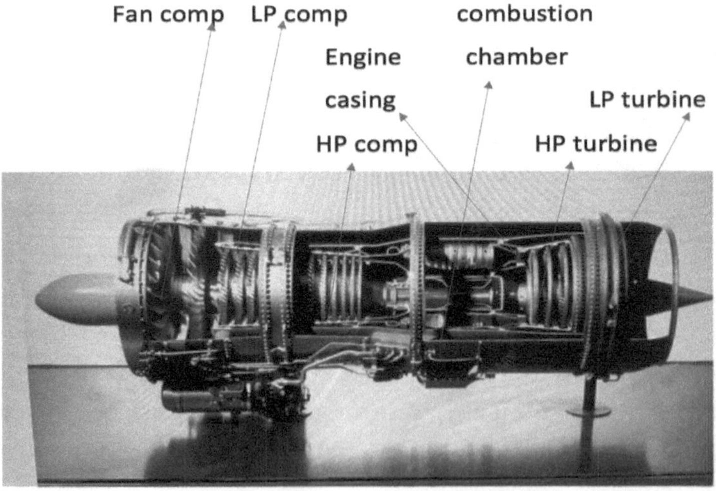

The basic major parts of the aeroengine remain the same in modern engines; however, the introduction of modern sensors, smart materials that can withstand higher turbine inlet temperatures, and the Full Authority Digital (Electronic) Engine Control (FADEC) system enhance the overall engine efficiency, fuel consumption, and reduce greenhouse gas emissions. The aeroengine can also be depicted in a more simplistic form, as shown below:

Fig 2.4: Line Diagram of Two Spool Aeroengine

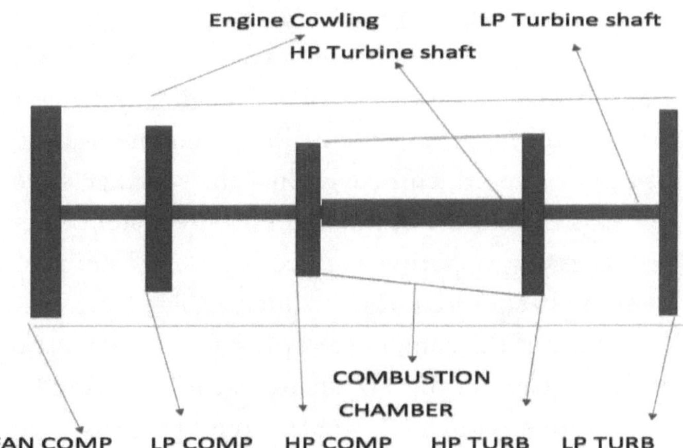

As explained earlier, in a simplistic form, the aeroengine comprises a fan compressor that inducts a large volume of air into the engine, compresses it, and raises the pressure of the air. In modern aeroengines, a large part of the air does not enter the core engine and exits at a higher speed from the rear of the engine, thus, providing thrust. The amount of air allowed to exit from the aeroengine without going through the core depends on the type of engine, i.e., a high bypass engine or a low bypass engine. In modern commercial aeroengines, the bypass air provides up to 80 per cent of the rated thrust. In contrast, for military fighter aircraft, low bypass aeroengines are used since the frontal area of the aeroengine cannot be large due to its installation inside the airframe itself.

In the arrangement of two spool engines, the fan blades, Low Pressure (LP) compressor, and LP turbine are placed on a long shaft known as the LP shaft, whereas the High Pressure (HP) compressor and the HP turbine are placed on another shaft known as the HP shaft. The LP shaft passes through the HP shaft, and a bearing is placed between these shafts, known as an inter-shaft bearing, and it is continuously lubricated under pressure. Two-spool or three-spool engines are designed primarily to obtain a higher air compression ratio for the engine with a smaller casing. It is pertinent to note that Rolls-Royce has successfully designed and developed three-spool aeroengines (Trent series) for large-scale utilisation on commercial aircraft.

There are more design complexities involved in multi-spool rotor design in terms of types of bearing, lubrication system design, providing core passages in the engine casing, oil scavenging system, heat transfer mechanism, and oil filtration system. The jet engine bearings generally work on the dry sump principle, and, thus, the scavenge system capacity is 2-3 times higher than the bearing oil pressure system. The aeroengine LP shaft is generally supported at three points, namely, the front (near the first or second stage compressor), intermediate (inter-shaft) bearing near the last stage of the compressor or inside the combustion chamber in the bearing slot inside the HP shaft, and rear bearing (between the HP and LP turbine segments), whereas the HP shaft is supported at two points, namely the front (near the last stage of the compressor) and rear bearing (near the first or second stage HP turbine). The inter-shaft bearing is provided mainly for providing spacing between the LP and HP compressor-turbine shafts, and it does not carry any load. The aeroengine shafts (LP and HP) on a typical modern low/high bypass aeroengine rotate between 8,000 and 15,000 Revolutions Per Minutes (RPMs) on civil as well as military aircraft. Hence, the support bearings and oil lubrication systems need to be designed meticulously to ensure optimum heat dissipation and continuous, uninterrupted bearing lubrication. In some of the aeroengines, an additional oil bottle is provided to ensure the smooth function of the engine under emergency conditions in the event of oil leakage in the main system or failure of the oil pump. It is important to note that most of the aggregates, such as Alternating Current (AC) and Direct Current (DC) generators, hydraulic pump main and standby, centrifugal breathers, etc., have a quill shaft with a pre-design weak point to get sheared off in the eventuality of their failure without affecting the aeroengine. In contrast, the oil pump shaft is a solid shaft with no provision to pre-design weak points to get sheared off, and it will keep running and provide lubrication to the support bearings as long as the engine is running.

The aeroengines use ball and cylindrical roller bearings as support, depending on the load function characteristics. The ball bearings can cater for radial as well as axial (thrust) loads, whereas the cylindrical roller bearings can be used for radial (thrust) loads only. Generally, on most multi-spool engines, the front bearings are ball bearings, and the rear bearings are roller bearings by design, with proper bearing housing to ensure minimal oil waste and an efficient scavenging system with a micronics filter and metal detectors (which can detect any ferrous material as debris in the oil systems). The fan blades in the bypass engines are much longer and wider compared to the pure jet engines, where the blades can induct large volumes of air through the air intake. The conventional stainless-steel blades will be much heavier, and, thus, in almost all types of modern bypass engines, the fan blades are made with titanium material; however, the recent trend includes the introduction of Ceramic Matrix Composite (CMC) blades. The designers generally provide a mid-segment contact point between the blades so that during rotation at high speed, these blades do not strike each other, even if there is some mild bending movement of the blades. The contact points are called snubbers by Safran, France, and clappers by Rolls-Royce, UK. The designers also recommend a hard material deposition, such as tungsten, on these contact points in order to avoid any wear during the normal operation of the aeroengine.

The LP compressor blades and HP compressor blades are made either from steel or titanium alloy, and the compressor discs are generally made from titanium alloys. The combustion chamber in most modern engines is either annular or can-annular, and made from nickel alloys, since the temperature inside the combustion chamber can go up to 2500°C, i.e., much above the melting temperature of the base material; however, the air cooling system is specifically designed to cool the combustor surface liners from the upper and lower channels by mixing the cool air from the bypass air channel with the core engine compressed air through small orifices all over the circumferential rim of the combustor body. It is important to note that only 25 per cent of the compressed air entering the combustion chamber is used for combustion, whereas 75 per cent of the compressed air is used for the cooling of the combustion chamber's inner liners. The combustion chambers may be designed in different configurations, such as annular or can-annular; however, the design caters to less than one-third of the air entering the combustion chamber that will be mixing the fuel for combustion purposes. In the annular combustion chamber, the burners have a lower surface-to-volume ratio in comparison to can-annular burners, and, thus, they require less cooling air. The burner weight is also comparatively lower in the annular combustion chamber than in the can-annular combustion chambers.

Fig 2.5: Annular Combustion Chamber

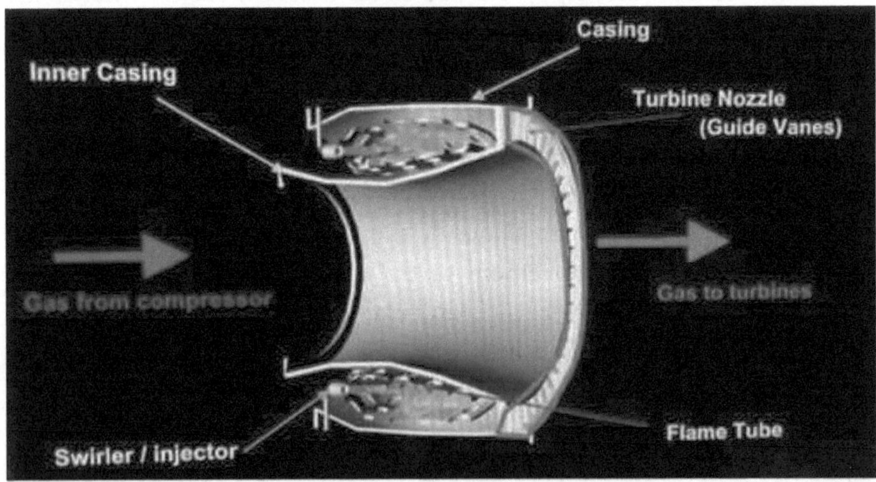

It is important for the designers to ensure the flame elongation length and ideal velocity (approximately 4-8 m/s) of the air in the combustor to ensure flame stabilisation so that the flame does not get extinguished during operation of the engine at any of the engine operating ratings. Generally, the velocity of air entering the combustion chamber is approximately 150 m/s in most jet engines; however, the kerosene flame rate is approximately 9.14 m/s. The fuel is mixed with the compressed air in an ideal ratio of 1:15, with the flexibility to operate normally from 1:10 to 1:20 by weight. The lean or rich air-fuel mixture will result in inefficient operation or even the flameout of the aeroengine. The role of automiser tubes, swirlers, and the location of ignitors are considered extremely important factors for fuel efficiency. Most of the designers use fuel-oil heat exchangers to transfer the hot oil heat from the engine oil to the fuel system so that the oil can be cooled and the fuel can be heated up to get atomised in the specially designed tubes and swirlers and sprayed in the combustion chamber to achieve better fuel efficiency.

Fig 2.6: Axial Compressor

The rotor compressor blades rotate inside the engine core casing, and the gap between the tip of the aeroengine compressor blades and the static casing should be zero or close to zero so that the air leaks from these gaps are minimal or negligible so as to ensure minimal compressed air pressure loss. It is necessary to take measurements of the shortest and longest blade lengths of each compressor stage, and the blades are subjected to tip grinding with reference to the shortest blades installed in each compressor stage. The static compressor casing for each stage is generally made up of a metallic semi-circular ring of titanium material or a specified shallow groove is designed in the LP compressor segment and nickel alloy in the HP compressor, which is coated with materials having abradable characteristics. The material characteristics should be such that it should be abradable, non-sparking, have a low brittleness factor, machinable, have excellent adherence to the base metal, and allow the rotating blades to make a groove during rotation without getting cracked or chipped off, thus, keeping the air gap minimal to enhance the compressor's efficiency. It is also essential that the blades be weighed during installation. The method of installation in the disc root is strictly followed as per the design criteria to ensure equal loading of the compressor disc, thereby minimising the risk of excessive centrifugal forces acting in any specific segment of the disc due to centrifugal forces generated at the blade during high-speed rotation. Most aeroengine manufacturers follow the general rule of installing equivalent compression disc loading patterns or equal and opposite weight rules of compressor and turbine blade fitment for equal disc loading. For example, if the heaviest blade is installed at a particular point on the disc, the next heaviest blade should be installed 180 degrees opposite the heaviest blade, and so on, either clockwise or anti-clockwise, on the entire circumference of the disc.

The compressor disc, prior to installation on the engine, must be balanced either by removal of material from the specified locations or by adding additional weight along with the blade in the blade bracket on the compressor disc at the specified imbalance point location. The compressor must be balanced at least thrice with a minimum of two or, preferably, three consecutive identical readings on the balancing machine in each quadrant. Each compressor blade must be locked at the root either individually through a locking strip (Russian pattern) in the blade root on the compressor disc drum circumference and bent towards the rear of the individual stage disc or through a circular locking spring strip in the rear stage passing through each blade to retain the individual blades at the specified location (Western pattern) firmly.

As mentioned earlier, the combustion chamber used on most modern bypass aeroengines is either annular or can-annular. The modern combustors are the most efficient major part of the aeroengine and generally operate at close to 100 per cent adiabatic efficiency. The material for the combustor is normally nickel-dominated superalloys. The fine cooling orifices are provided all around the circumference for a larger part of the compressed air entering the combustor for continuous cooling. The combustion chamber's outer surface is also temperature-controlled for active-clearance control of the aeroengine. The material should be machinable and weldable for crack repair during the scheduled repair or overhaul and be able to withstand high thermal stresses caused by the high working temperature of the hot gases.

Fig 2.7: Impulse-Reaction Turbine

It is well known that while the turbine inlet temperature is increased, the thrust of the engine increases significantly; however, the air gap between the rotating turbine blades and the static engine casing should be maintained as low as possible so as to reduce the hot gas leakage in order to improve engine efficiency. This is achieved by the placement of a metallic honeycomb circular ring either in a single piece or, preferably, in several segments that are fixed to the turbine casing slots. Unlike the compressor blades, the material of the turbine blades should be able to withstand these temperature regimes for prolonged operations.

The turbines used in jet engines can be either impulse turbines or reaction turbines, or a combination of these two. In modern bypass engines, generally, most designers use impulse-reaction turbines to obtain the benefit of impulse and reaction characteristics. The turbine blades, during the operation of the engine, are subjected to a very high degree of thermal stress and turbine inlet temperatures up to 1400°C, and roughly each turbine blade is expected to handle approximately two tons of air continuously and produce work from the high temperature, high-velocity gases impinging on the turbine blades. The turbine blades are made from either nickel alloys or cobalt alloys in order to withstand high temperatures and high thermal stresses. Another important factor in turbine blade metal selection consideration is high fatigue strength and low creep value for prolonged periods of operation. The turbine blades are normally hollow, and the cooling holes are provided with designed holes in the trailing end wherein the cool air from the bypass air is made to flow from the turbine blade root with a pre-designed cavity inside the blade for equal distribution of airflow to ensure continuous cooling of the turbine blade surfaces. The thermal barrier coatings on the turbine blades and the Nozzle Guide Vanes (NGVs) have to withstand high temperatures to the tune of 1000°C and above. A special coating material is used, which is an aluminium-silica-based alloy named yttria, ytterbia, or Yttria Stabilised Zirconia (YSZ), porous ceramics, or the alloy of softer ceramic metals with a base coat and a harder material as a top coat with different characteristics. YSZ is the most popular thermal barrier coating, generally coated with plasma spray machines or chemical vapour deposition on the external surface of the turbine blades. It provides wear and corrosion resistance, besides withstanding thermal stresses. There are some other specialised metal powders available for use as bond coats and top coats for use as Thermal Barrier Coatings (TBC); however, the suitability of the material for the base metal's adherence and wear characteristics needs to be studied carefully.

Fig 2.8: Low Pressure Turbine

It is important to note that there are a number of coating and wear-resistant materials for corrosion protection slurry available for aeroengine parts, such as Sermeloy-J, Serme-T, and Sermetal-W, which are already in use. However, the selection is based on the metal characteristics and chemical compositions. The coating material should have the specific characteristics of good galvanic corrosion resistance and oxidation resistance, at least up to 1000ºC. The turbine disc and the blades are subjected to cyclic heating and cooling, thereby causing high functional fatigue besides various forces acting on them and causing additional stresses. The size of the turbine blades from the first stage to the subsequent stages keeps increasing since the hot gas' pressure reduces. Still, the volume increases, and, hence, a larger turbine blade contact surface area is required to extract power from the relatively lower pressure and high volume of gases. The LP turbine blades of gas turbines, as well as steam turbines, are generally shrouded in order to avoid the extremely dangerous effect of the resonance frequency. We are aware that if the speed increases and the frequency remain constant, the wavelength will increase as per the equation: $\lambda = v/f$ (λ is the wavelength, v is the velocity, and f is the frequency of the object). The military aircraft aeroengines are also incorporated with an afterburner mechanism to burn additional fuel in the jet pipe to raise the gas temperature and its volume and obtain much higher jet velocity. It is possible to achieve up to 50 per cent additional thrust, and, hence, it helps in the reduction of the aeroengine size for military aircraft.

The turbine discs are made from nickel or cobalt-dominated steel superalloys, with specific characteristics such as high tensile strength, low creep, fatigue resistance, corrosion resistance, etc. The turbine disc and blades are the most stressed parts in the engine as they have to operate in the higher envelope of the material thermal range, produce high power in terms of thrusts, and propel the aircraft, besides providing electrical and hydraulic power for operating various services and controls of the aircraft. The turbine blades are generally designed as hollow blades with an internal core cavity in order to ensure high-velocity air cooling provisions. The cool compressed air enters the blades from the root and exits from the trailing end, thus, transferring the heat at a specified, designed rate.

The hot gas velocity exiting from the first stage of turbine blades enters the nozzle guide vanes. It impinges on the subsequent turbine stages in the multi-stage turbine engine at the correct angle, and, thus, the process continues till the hot gas exits from the last stage of the turbine. The hot gases, prior to mixing with the environmental air, are made to pass through a specifically designed jet nozzle configured to reduce the noise and swirling actions. In military engines, an afterburner system is designed and incorporated into the jet pipe to burn additional fuel in the jet pipe through an additional burner ring in order to produce additional thrust of up to 50 per cent. Thus, the exhaust air from the core engine mixed with a part of the bypass air is further heated to increase the temperature of the gases and made to exit from the jet nozzle with much higher velocity to obtain the higher resultant thrust.

The different types of aeroengines most widely used on military and civil aircraft and helicopters are covered below.

Turbojet

These are the initial types of gas turbine engines developed; they follow the same principle of operation as the reciprocating engine through the cycle of operation as air intake, compression, power, and exhaust; however, all four actions are performed concurrently in the jet engine, whereas in the reciprocating engine, it is one action at a time followed by another. Hence, for the same size of reciprocating and jet engines, the power generation from the jet engine will be higher.

Turboprop

Turboprop engines are quite versatile and are generally used for slow-speed aircraft. The turboprop engines is basically a jet engine, wherein a gearbox is used to extract the power of the jet engine to rotate the propellers. A power shaft through a gear arrangement is used to reduce the rotational power of the main engine shaft through a sun and planetary gear system in single or multiple stages. The propellers are mounted on the shaft originating from the reduction gear.

Turboshaft

Turboshaft engines are used on helicopters. The power generated by the jet engine is extracted through a bevel gear arrangement and taken to the main gearbox. The RPM of the engine is reduced manifold through the sun and planetary gear system in single or multiple stages and utilised for rotating the main rotors and tail rotors.

Turbofan

Turbofan engines are the type of jet engine with comparatively bigger frontal openings and higher chord (width) first-stage compressor blades, and post-compression, a part of the air is directly sent to the exhaust cone, bypassing the core engine. However, another part of the compressed air goes inside the core engine for normal operation, as per the typical jet engine. The amount of air that passes through the core engine and the exhaust duct depends on the Bypass Ratio (BPR). Generally, the turbofan engines used for fighter combat aircraft utilise low bypass engines since their frontal opening will remain low, and these are installed inside the aircraft structure due to higher agility and manoeuvrability requirements. However, for transport aircraft engines, which are strapped under the wing, the engine frontal opening is much higher, and, thus, high bypass engines are used, wherein up to 80 per cent of the thrust is produced from the bypass air. Modern aeroengines for civil and commercial aircraft have a BPR up to 12:1; but for military fighter combat aircraft, the BPR is much lower and generally in the range of 0.2:1 to 1.5:1. However, for high-performance fighter aircraft with speeds above 2 Mach, the BPR will remain in the lower limit, i.e., approximately 0.2:1 to 0.5:1. It is understood that the P&W-made F-135 aeroengine installed on the F-35 combat aircraft has been designed for a BPR of 0.4:1. The BPR of the GE F110 aeroengine, used to power the F-16 aircraft, is 0.87:1. The BPR of CFM International's new LEAP turbofan engine is 9:1. In contrast, the PW1100G series designed to fly the A320 NEO family has the highest BPR of 12:1, which is considered the highest in aviation history.

3

Historical Perspective

India was practising an absolutely independent economy prior to British colonial rule, wherein it had a large export base for handicraft items, cotton, silk textiles, spices, metal, and precious stone works, with a sizeable share of world trade. The Indian economy underwent turmoil during the British rule, which ensured that the raw materials from India were exported to England, and India was made a mere consumer of the finished goods for the industrial products produced in England. In fact, India could not achieve the desired benefits of the industrial revolution in Europe.

In the aviation segment also, our initial orientation for the design and development of aircraft or the aeroengine was non-existent, primarily due to the low industrial base in India, the denial of technological access from the industrially advanced countries, and a lack of a technological fundamental research base. All the aircraft in the civil and military domains procured in India were from foreign-based OEMs. Thus, the option of manufacturing aircraft or aeroengines in India was not there. The level of industrial base in the country and the technical knowhow for the purpose of design, development, and manufacturing of aircraft or aeroengines were inadequate post-independence. The government primarily concentrated its focus on the core sectors such as agriculture, and limited industrial sectors such as steel, power plants, cement, fertilisers, roads, and other infrastructural projects.

The design, development, and manufacturing of the HF-24 Marut fighter aircraft by Hindustan Aeronautics Limited (HAL), Bangalore, in the 1960s, as authorised by the Government of India, was, indeed, a first and a giant step in the aviation history of India. Although HAL had employed a German engineer, Kurt Tank, as a design consultant, the complete fabrication, assembly, integration, and testing of the aircraft were done in-house. The HF-24 Marut, the first Asian fighter aircraft, undertook its maiden flight on June 17, 1961. The aircraft went into series production, and the first aircraft was delivered to the Indian Air Force (IAF) on April 1, 1967. HAL produced a total of 147 Marut aircraft, mainly for the IAF. The aircraft was extensively utilised for ground attack roles and participated in combat action during the 1971 India-Pakistan War. Although the aircraft was designed for supersonic performance and an interception role, it was due to the inadequate power of the aeroengine that the aircraft did not undertake the specific role of interception during the operational exploitation phase and was made to retire prematurely in 1982.

It is learned that pursuant to the induction of indigenous fighter aircraft in India, some East Asian countries expressed their desire to procure the HF-24 Marut fighter aircraft; however, the same was not agreed upon by the Government of India due to the policy of non-export of military weapons to any foreign country. The indigenous production of the HF-24 Marut and its induction into the IAF, even with a limited role or success, was indeed a great step.

The lessons learnt from the successful manufacturing of fighter jets by HAL for the first time in Asia, the operational exploitation of indigenous aircraft by the IAF, including combat missions during the Indo-Pak War in 1971, the export denial and sale of combat fighter aircraft to foreign friendly countries, the premature retirement of the aircraft from the IAF due to limited exploitation, and sustainability issues due to the inadequate power of the aeroengine were not given due importance. The corrective measures were not instituted by the concerned stakeholder groups in a systematic and chronological order. It is understood that neither was indigenous aeroengine design and development initiated nor was alternate aeroengine procurement from abroad considered for the sustenance of the Marut indigenous combat aircraft. Thus, the indigenous design and development initiative and capability of HAL ended for a long time until the LCA design and development got restarted. Once again, it is feared that the LCA-II or the futuristic AMCA may have a similar fate if indigenous aeroengine development is not given due priority in the near future, since the GE-414 proposed for installation on LCA-II has a power limitation of 98 KN against the requirement of 110 KN. GE may demonstrate temporary or short-term thrusts of 100-105 KN by enhancing the fuel input; however, the same cannot be sustained for prolonged operational usage by the IAF. Therefore, it is imperative to learn from history, and all-out efforts should be made to design and develop an indigenous, low bypass turbofan core aeroengine of 110-130 KN thrust for military combat aircraft as well as single-aisle civil aircraft applications, with minimal changes.

Notwithstanding the historical perspective, the Government of India decided to go ahead with the design, development, and manufacturing of the indigenous LCA in August 1983 at a cost of ₹560 crore, with a time span of 8-10 years. It established the Aeronautical Development Agency (ADA) in Bangalore in June 1984 as per the provisions of the Societies Registration Act, 1860 under the Ministry of Defence, with an exclusive role of LCA project management. The IAF issued the Air Staff Requirement (ASR) in October 1985 with the aim of procuring 220 aircraft (including 20 trainer aircraft). The project was divided into three phases in the format of Full-Scale Engineering Development (FSED) for the purpose of faster development and monitoring at various levels. Subsequently, the Government of India sanctioned ₹2,188 crore for FSED-I in February 2000. Further, for the next phase, the Government of India again sanctioned ₹2,431.55 crore for FSED-III in November 2009. The development of certain critical systems of the aircraft, such as the aeroengine, Multi-Mode Radar (MMR), Self-Protection Jammer (SPJ), and Radar Warning Receiver (RWR), could not be developed due to the complexities of the technology. ADA decided to go for direct procurement of these aggregates and systems from foreign vendors, for which another financial sanction was accorded by the Government of India. A total of ₹10,397.14 crore was sanctioned by the Government of India for LCA development in various phases, excluding the cost of the Kaveri engine development.

The development of the LCA got delayed considerably due to numerous reasons. The LCA got initial operational clearance in December 2013, after a lapse of 30 years. As per the original project report, full operational clearance was scheduled to be issued by December 2008. The IAF had planned to replace the MiG-21 variants, manufactured between 1966 and 1987, with the indigenous LCA. However, due to the long delay in the developmental process, the IAF was compelled to opt for the upgrade of 125 MiG-21 aircraft and also enhance the life of other MiG-21 variants to ensure fleet strength and availability.

The responsibility for the design, development, and manufacture of suitable aeroengines for the LCA was entrusted to GTRE, Bangalore. It is important to note that pursuant to the Government of India's sanction for LCA development, ADA, along with HAL and GTRE, carried out a feasibility study on the suitability of the GTX-37 engine already being developed by GTRE for use on the LCA. The study team agreed that the GTX-37, with the incorporation of a few changes, could be developed. Accordingly, the study team sent its recommendation to the Government of India.

Based on the feasibility study undertaken by ADA, HAL and GTRE for the possible use of the GTX-37 aeroengine with the incorporation of a few changes already under development by GTRE, the Kaveri aeroengine design and the developmental project were sanctioned by the Government of India in 1989, with an estimated cost of ₹382.81 crore, and the initial project completion date in December 1996. The project got severely delayed due to the complexities involved in the aeroengine design and development. The project cost, as well as the time-frame, was revised several times up to December 2009. The revised cost was ₹2,839 crore, with ₹2,105 crore [the Foreign Exchange (FE) element of ₹1,227 crore] earmarked for interim flight trials with the LCA and the remaining ₹734 crore for the final flight standard engine. GTRE had progressed the Kaveri aeroengine development and produced nine full prototype engines and four core engines. A total of 2,882 hours of engine testing was conducted, and the aeroengine was also sent to CIAM in Russia for flight testing, wherein the aeroengine performance was observed to be at variance from the specified parameters. Thus, the aeroengine was not cleared for exploitation on the LCA.

It is understood that GTRE is presently attempting to review the Kaveri aeroengine design or re-design the aeroengine of 110-130 KN thrust for the LCA Tejas, and AMCA, with the help of foreign-based OEMs such as Safran, France, Rolls-Royce, UK, or GE, USA. It is expected that GTRE may enter into a joint venture with foreign vendors for the future development of suitable engines. However, it is pertinent to note that none of the foreign-based original equipment manufacturers will part with the critical design data and will only enter into the transfer of manufacturing technology. In order to co-design the aeroengine, GTRE, Bangalore, may enter into a research and development agreement from scratch and co-share the Intellectual Property Rights (IPRs) with foreign original equipment manufacturers, subject to the willingness of both parties. However, this aspect will be extremely expensive and time-consuming. It is considered prudent to opt for collaboration with the foreign OEM only in those specific systems or modules of the Kaveri aeroengine design and development that are less reliable, less efficient, and probably failure-prone or where the technical knowhow at GTRE is not considered adequate.

4

MRO of Aeroengines in India

The MRO of aeroengines is performed at specified intervals based on the flying hours done by the aeroengine or on a calendar basis, i.e., the time from the date of manufacture of the aeroengine or a specified time-frame elapsed after the last MRO. Some of the aeroengines are also required to be withdrawn from service based on the cycle time, i.e., one start and switch off is considered as one cycle since it is considered that the aeroengine undergoes a thermal shock each time it is started. Sometimes, a military combat aeroengine is also withdrawn from service if the aeroengine's peak performance time-frame increases beyond the specified limit since it is amply clear that the aeroengine undergoes the maximum thermal stress during the peak performance operation, i.e., afterburner engagement.

Modularity of Aeroengine and Scope of MRO

In modular aeroengines, the life of each module is specified separately based on the material, technology, and engineering processes used during manufacturing and the stresses the specific module undergoes during the exploitation of the engines based on the static or dynamic characteristics of the loading pattern. Sometimes, the aeroengines are also withdrawn from service mandatorily due to bird ingestion or foreign object damage, depending on the severity of the damage. As per Mordor intelligence, the commercial aeroengine MRO market is likely to grow at a Compounded Annual Growth Rate (CAGR) of

6.37 per cent from US$ 37.75 billion in 2023 to US$ 51.41 billion in 2028. It is learnt that since commercial aircraft flying got severely restricted in most parts of the world due to the COVID pandemic in 2020, the aeroengine MRO requirement also reduced drastically in the year 2021; however, since mid-2022, commercial aircraft operations have resumed in a normal manner, and it is expected that the MRO requirement for aeroengines will increase significantly.

The MRO market is mainly segmented into three parts: commercial airlines, military aviation, and general aviation. Generally, the commercial aviation market captures more than 80 per cent of the MRO market share approximately, whereas the military and general aviation markets are approximately 10 per cent each. Thus, the commercial aviation market share is likely to dominate the MRO business due to the greater number of commercial aircraft compared to military aviation and the high cost of aeroengine maintenance compared to general aviation. In the recent past, the Asia-Pacific region has added a large number of aircraft as part of its fleet expansion in order to cater for a higher number of passengers in the near future. In view of this development, most of the MRO companies, including aeroengine OEMs in the USA and Europe, are establishing MRO facilities in China and India. Air China announced its joint venture company with Rolls-Royce in September 2022 for the MRO of Rolls-Royce aeroengines such as the Trent 700, Trent XWB-84, and Trent-1000. Both companies will hold 50 per cent of the shares in the joint venture company, worth US$ 378.2 million. French aeroengine manufacturer Safran also claims to have started an MRO facility in Suzhou, China, in February 2022.

In view of a large fleet expansion by Indian airline operators such as Air India, Indigo, and other commercial airlines, Safran, in July 2023, announced the decision to set up its biggest MRO facility in Hyderabad, India, with an investment in US$ 200 million. The facility would be capable of undertaking up to 300 aeroengine shop visits annually. The MRO facility would mainly cater to the CFM56, Leap 1A, and Leap 1B aeroengines, which are more commonly used on Indian single-aisle aircraft. CFM International dominates the single-aisle aircraft market globally with approximately 60 per cent of the market share. In India, approximately 75 per cent of single-aisle aircraft are powered by CFM international aeroengines only. It is predicted that the aeroengine MRO segment will remain the largest and fastest-growing sector of the commercial aviation jet MRO business over the next decade. The prominent global MRO companies are as follows:

(a)   Lufthansa Technik
(b)   Rolls-Royce Holding PLC
(c)   General Electric Company
(d)   Safran SA
(e)   P&W
(f)   Raytheon Technologies Corporation
(g)   Singapore Technologies Engineering Ltd.
(h)   MTU Aeroengines
(i)   Delta Airlines Inc.
(j)   Hong Kong Aircraft Engineering Company Ltd.
(k)   Textron Inc.
(l)   Honeywell International Inc.
(m) S7 Technics
(n)   SR Technics Switzerland Ltd
(o)   Lockheed Martin Corporation

In India, the MRO of commercial airline aeroengines is carried out by the following companies:

(a)   Air India Engineering Services Limited (AIESL)
(b)   Air Works
(c)   Indamer Pvt Ltd
(d)   Deccan Charter
(e)   Taj Air
(f)   Bird Execujet
(g)   GMR Aero Technic Ltd
(h)   Max MRO Pvt Ltd

AIESL is a subsidiary company of Air India. AIESL is the only Indian company to have a full-fledged engine overhaul facility at Mumbai and Delhi airports. Other Indian companies presently provide only limited MRO services.

The military aircraft aeroengine MRO in India is performed by HAL and the integral base repair depots of the IAF. HAL has been engaged in the licensed manufacturing of a large range of aeroengines, as enumerated below:

HAL (Bengaluru Division)
(a)   Rolls-Royce engine Adour 811/804M2 (Jaguar aircraft)
(b)   Rolls-Royce Adour 871-07 (Hawk aircraft)
(c)   TM 333 2B2/TM3332M2 (ALH MK I&II/Cheetal helicopters)
(d)   Shakti 1H1
(e)   Artouste IIIB/B1 (Chetak/Cheeta helicopters)
(f)   Garret TPE 331-5 (Dornier aircraft)
(g)   Dart 533-2 and 536-2T (Avro aircraft)
(h)   Orpheus 70105 (Kiran Mk-II aircraft)
(i)   Gnome 1400-1T (Seaking)
(j)   GTSU-110 (LCA)
(k)   Amagb (LCA)
(l)   Pilotless Target Aircraft Engine (PTAE) Lakshya

HAL (Koraput Division)
(a)   AL-31 (SU-30 aircraft)
(b)   RD-33 (MiG-29 aircraft)
(c)   R-25 (MiG-21 aircraft)
(d)   R-29 (MiG-23/27 aircraft)
(e)   TG-16 M (APU for AN-32 aircraft)

Base Repair Depots of Indian Air Force
(a)   M53 P2 (Mirage-2000 aircraft)
(b)   Viper 22-8 (Kiran trainer aircraft)
(c)   AI-20 (AN-32 transport aircraft)

(d)   TV-3 MT (Mi-17 helicopters)
(e)   VK-2500 (Mi-17 V-5 helicopters)
(f)   AI-9V (Mi-17 APU)
(g)   Safir (Mi-17 V-5 APU)

HAL has been involved in the licensed manufacturing of aeroengines of Western origin, such as the Adour 811 and 804 M1, and Russian-origin aeroengines R-25, R-29, RD-33, and AL31FP, for more than 30 years. All the infrastructure in terms of machinery and Transfer of Technology (ToT) was provided by the OEMs, and the technology adoption and absorption were done by HAL in a well-coordinated manner. Generally, the technology transfer was done in a phased manner wherein, in the first phase, all the modules and aggregates were supplied by the OEMs, and only the assembly of the aeroengine took place in HAL. Subsequently, the raw material was supplied by the OEM, and the simpler parts were manufactured by HAL. Gradually, in the subsequent phases, the OEM provides most of the manufacturing drawings and the raw materials, and the complete manufacturing work is undertaken by HAL. However, the OEM does not provide design data, material specification data, processing technology, or forging or casting drawings. It is also important to note that some of the critical parts, such as the technology for the compressor and turbine blades, the fuel pump and control units, the hydraulic pump, the gearbox, the Full Authority Digital (Electronic) Engine Control (FADEC) system, etc., are supplied by the OEMs. Thus, the dependence on the supply of these spares continues for the entire service life of the aeroengine.

HAL has extensive experience in the licensed manufacturing of aeroengine parts, machinery, and technical acumen. It is also capable of undertaking any assigned task for the indigenous design, development, and prototype manufacturing of aeroengines. HAL and the base repair depots have been involved in the MRO of all types of military engines, such as turbojet, turbofan, turboprop, and turboshaft engines, for the last four decades.

The MRO task of aeroengine manufacturing differs from the new aeroengine manufacturing since, for the new engines, all the parts and components are new and, thus, conform to the engineering drawing. In contrast, for the MRO, several dynamic parts undergo wear and tear and, thus, need rework to bring these parts as per the specifications, or the rejected parts are replaced with new parts. Inspection and micrometry of the new parts generally do not lead to rejection, whereas the parts in the overhaul processes often get rejected and subjected to rework if the technology is provided by the OEM; otherwise, the parts get rejected and replaced with new parts.

As mentioned earlier, the aeroengines or their modules are inducted for MRO based on the following criteria:

- On completion of specified hours of operation since the first installation on the aircraft or after the last MRO.
- On completion of specified peak performance hours (operation of the afterburner or reheat-engagement).
- If the damage inside the aeroengine is ascertained to be extensive due to suction of Foreign Object Damage (FOD) or a bird strike near the air intake vicinity or sucked inside the engine.
- If the specified calendar life of the aeroengine has expired in storage and the storage servicing has not been performed as specified.
- If the aircraft has performed a high 'G' operation beyond the specified values or a heavy landing beyond the laid down limits.

MRO Process

A classical MRO process of the aeroengine involves the following steps:

Documents Perusal and External Inspection

The aeroengine documents are perused in order to list any specified operating parameters exceeded, incidents, accidents, failure of any module during operation, flame out, and modules or aggregates falling due for MRO or replacement. Primarily, the study of documents dictates the quantum of work involved and the likely time-frame it might require, besides the rough estimate of the MRO cost. The external physical inspection reveals if there is any sign of corrosion, especially if the aeroengine has been received from the coastal bases of operation and not properly preserved prior to packaging in the engine case. The external inspection also reveals if there was any external damage during packing, unpacking, or transit.

Dismantling of the Aeroengine

The dismantling of the aeroengine is done as per the overhaul or repair manual, and some of the critical measurements, such as turbine disc creep or elongation measurement, are done to compare with the original values and check if any deformation has taken place, which may not be visually noticeable.

Cleaning of Parts, Assemblies, and Sub-Assemblies

All the aeroengine parts are thoroughly cleaned with the various cleaning agents specified by the OEM overhaul manual, depending on the basic material of the parts. Since the cleaning agents may be acidic, alkaline, or extremely poisonous compounds, the parts are properly cleaned by running hot and cold water subsequently so as to avoid any corrosion due to residual cleaning agents on the parts. Some of the aeroengine parts, such as the oil pump or fuel pump, are painted externally; the paint is to be removed through sandblasting or grit blasting, depending on the material of the pump, and once again, the pump is subjected to thorough flushing and cleaning. All the aeroengine parts, aggregates, and modules are sent to their respective workstations for further disassembly to the smallest components.

Inspection and Micrometry

All the aeroengine parts and aggregates are subjected to visual and optical inspection. With 10x glass as per the overhaul or repair manual for any deviation, the micrometric measurement is performed to ascertain if it is conforming to the specified limits of various laid-down parameters. If all the parameters of the components are in conformance with the specified values, the component is cleared for further assembly, or else the overhaul manual is referred to for the relevant technology for repair and reclamation. The parts are repaired and once again subjected to micrometric measurement. If the component passes through the laid down checks, it is cleared for further assembly, or else it gets rejected, and these parts need to be replaced with new ones. Some of the mechanical parts are also subjected to fluorescent penetrant inspection, red-dye inspection, magnetic flux testing, X-ray testing, and other non-destructive tests to detect any minute cracks or internal deformities. Generally, mechanical parts with the slightest cracks are replaced with new parts for reliability and safety. If the mechanical parts pass the relevant check, they are cleared for further assembly. Some of the parts, such as turbine blades, are also subjected to vibro-tumbling to create a smooth, polished, and shiny surface. The turbine disc fir tree slots where the turbine blades are retained are also one of the most stressed and fatigued parts, and, hence, they are also subjected to shot peening to improve fatigue strength and compressive residual stresses.

Assembly of Components into the Aggregates/Modules

Pursuant to inspection and micrometric examination, the sub-parts and components are assembled as per the overhaul or repair manual, the aggregates and modules are subjected to specific test rigs for comprehensive testing, and the results are evaluated against the laid-down parameters. The aggregates are cleared for further assembly on the aeroengine; otherwise, they are subjected to re-repair or a minor adjustment of controls. Some of the major assemblies such as low-pressure compressors and turbines and high-pressure compressors and turbines are subjected to balancing. The compressor and turbines are first individually balanced, then mounted with the respective turbines and re-balanced. In order to balance them, the compressor and turbines are run at approximately 10-15 per cent RPM since the bearing lubrication system is not available on the balancing machines. However, if the compressor or turbine is balanced well at 10-15 per cent of the rated RPM, it generally remains balanced at the rated RPM as well. The addition of weight is normally done in the blade slots or at designated places on the compressor or turbine discs. If some weight is to be removed, the same is to be done at the earmarked places. The balancing is to be done multiple times, wherein the same reading must be repeated at least twice or preferably thrice.

Testing of the Individual Aggregates

It is essential to carry out rig testing of all the aggregates on their specific test rig to check their performance as per their test protocol. The actual values are annotated in the aggregate test certificate, and the values are checked as per the overhaul/repair manual and also with respect to the last rig test values to ascertain any performance degradation. The electrical looms are checked for continuity and insulation and repaired in accordance with the laid-down procedure. All the sensors are also checked for their optimum performance and re-installed if they pass the relevant checks as per the testing procedure.

Final Assembly of the Engine

All the modules and aggregates, after rig testing and inspection, are rooted to the final assembly section, where the engine is initially assembled in the vertical position wherein the HP compressor and turbine are installed with the HP shaft passing through the combustion chamber, and through the HP rotor, the LP turbine attached with the shaft passes through, and the LP compressor is assembled from the front side. Subsequently, the aeroengine takes the horizontal position for strapping the various aggregates, such as the fuel pump, afterburner fuel pump (on military engines), hydraulic pump, Alternating Current (AC) and Direct Current (DC) generator, and the gearbox. The electrical looms are connected to various points, and the sensors are installed. The aeroengine post assembly is checked by an independent team not related to the assembly team to check the correctness of the workmanship in accordance with the standard test protocol checklist. The documentation of the aeroengine is carried out, and all the parts, aggregates, and modules installed on the aeroengine should preferably have full overhaul life so that the aeroengine need not be withdrawn from service for replacement of the parts, aggregates, and modules.

Ground Test-Bed Run

The aeroengine post assembly is subjected to the ground test-bed run, wherein all the tests are carried out as per the test protocol, and the various parameters are noted in the test protocol and the aeroengine log books. All the parameters must conform to the laid down specification, and, if required, minor adjustments are performed to obtain the specified values. The aeroengine is certified fit for operational usage for specified hours before the next visit to the overhaul workshop. If any module or the aggregate does not have a specified residual life equivalent to the MRO life of the aeroengine, the same is to be annotated clearly as caution.

Documentation, Preservation and Packing

The aeroengine, once tested on the ground test-bed successfully, is preserved and packed in the specified engine case. The preservation must be done as per the specified procedure so as to prevent any corrosion in the internal part of the aeroengine and even on the outer surface of the aeroengine. It is mandatory to endorse all the major work done on the aeroengine during the MRO process, as well as the engine performance parameters as observed on the engine test-bed and the specified operational life of the various modules and sub-assemblies and all the aggregates in the aeroengine log book.

5

Current Status

India has been attempting to indigenously design and develop a suitable aeroengine for the LCA Tejas Mk-I, Mk-IA, Mk-II, and AMCA with an approximate thrust of 80 KN. GTRE, Bangalore, has been working relentlessly on the design, development, and prototype manufacturing of aeroengines for the last three decades but so far has not been successful in its developmental process. It is understood that GTRE is in the process of working with, and obtaining design assistance from, well-established, proven foreign-based aeroengine designers and manufacturers for the co-design of aeroengines.

As covered in the previous chapter, GTRE has been working for over three decades on the design, development, and prototype manufacturing of indigenous aeroengines and has spent approximately ₹2,100 crore on this. However, it did not succeed in the design and development of the Kaveri engine for installation on the LCA, and the probable prominent reasons, as covered earlier, are enumerated below:

(a) Inadequate dry and wet thrust of the engine against the laid-down specifications.

(b) Inadequate thrust-to-weight ratio, wherein the Kaveri aeroengine thrust-to-weight ratio is approximately 5.7 in comparison to the M-53P2 aeroengine of the Mirage-2000 ratio of 6.4 and the M-88 aeroengine of the Rafale aircraft ratio of 8.5.

(c) Excessive vibration of low-pressure turbine blades at higher altitudes.

(d)   Excessive noise generation at a resonant frequency.
(e)   The excess weight of the engine against the specified specification.

It is understood that GTRE, post-closure of the Kaveri project, is probably working on the design and development of marine engines for Indian Navy applications and also for unmanned flight vehicles.

It is also understood that the negotiations for co-design and development of a modular core aeroengine of suitable rating to the tune of approximately 110-130 KN are in the advanced stage with Safran, France, Rolls-Royce, UK, and GE, USA, for joint development in India. It is pertinent to note that the foreign-based OEMs may agree to part with the technologies for aeroengine manufacturing already developed by them for other aircraft programmes. However, this aspect will not be considered advantageous for an indigenously designed aeroengine in the absence of design, material specification, and special process data, thereby making India forever dependent on the OEM for any design alterations, upgrade modifications, life extensions, or reliability improvement plans in the future.

It is considered prudent to identify specific problem areas such as material selection, design for the compressor, combustor, turbine, aerodynamic characteristics, thermal, aerodynamic, and propulsive efficiency, FADEC, future noise and emission control mechanisms, incorporation of an afterburner system, and the diffusion process. It is also considered prudent to undertake an extensive study of the futuristic trend in aeroengine design in view of recent environmental challenges (air pollution and noise reduction) so that the aeroengine remains valid for adequate operational exploitation for at least 30-40 years.

Here, it becomes necessary to briefly discuss the contemporary state-of-the-art technologies pertaining to steam and wind turbine technologies in India. Even though the principles of operation of steam and wind turbines are at variance from the gas turbine engines, some of the industrial processes and machinery are applicable to gas turbine engines. Presently, there is a large number of industrial steam turbines being used in thermal power plants, nuclear power plants, cement industries, oil and gas, petrochemical refineries, and fertiliser industries in India. The steam turbines of higher power (more than 100 MW) are being manufactured by Bharat Heavy Electricals Limited (BHEL) in the public sector, and the lower power turbines (30-100 MW) are manufactured by Triveni Turbine, Bengaluru, and some other Indian companies in the private sector. Triveni Turbine is the market leader in power turbine manufacturing below the 30 MW segment, with a 55 per cent market share. It is important to note that the principles of operation of steam turbines and air compressor turbines are at variance; however, some of the technologies for the steam and aeroengine technologies, such as the manufacturing of blades, including machining, blade balancing, thermal barrier coating, special processes and treatments, dynamic bearings and oil lubrication system, gearbox, oil pump, fuel pump, stainless steel tube bending processes, and vibration characteristics, are similar. The technologies pertaining to non-destructive tests of various components of engines, disc creep and elongation checks, compressive stress checks, and residual stress checks of turbine disc materials are also similar. It is prudent to note that the technologies for steam turbines were procured by BHEL and Triveni Turbine from Siemens, Germany, and GE, USA, respectively. These companies have been manufacturing steam turbines and improving the various interrelated processes and quality aspects continuously.

It is also important to note that Triveni Turbine and General Electric (GE), along with Baker Hughes (BH), a subsidiary of GE, USA, formed a joint venture for steam turbine design, manufacturing, and sale in India on April 16, 2010. The company was named General Electric Triveni Limited (GETL), and the products were marketed as 'GE Triveni.' GETL was primarily formed for the design, development, and manufacturing of steam turbines with 30-100 MW capacity with the help of GE's technology. The two companies went into disagreement subsequently, primarily over the market access policy of GE. GE was very restrictive in allowing the joint venture company to open market access for its joint ventures. e.g., there was an inquiry from Union Flacq Power

Plant of Mauritius, wherein GE, USA, instead of redirecting the business to GETL, preferred to bid for the project itself directly. Moreover, GE, USA, acquired another company, ALSTOM, in India, which became a direct competitor to its own joint venture company with Triveni Turbine, GETL. Triveni Turbine also alleged that GE did not provide the requisite technologies as per the provisions outlined in the joint venture agreement. As such, it caused financial losses to the joint venture company. The case was filed in the National Company Law Tribunal (NCLT), Bengaluru, on July 1, 2019; however, subsequently, the case was settled by the parties on September 6, 2021, agreeing to the following terms:

- Termination of the joint venture.
- GETL, post-termination, to change the name and become a wholly subsidiary company of Triveni Turbine Limited.
- All intellectual property rights licensed by BH parties or TTL to GETL to be returned to the respective licensor.

A settlement consideration of ₹208 crore was transferred by General Electric, USA, through a BH subsidiary company, 'DI,' to Triveni Turbine Limited, Bengaluru.

Lessons Learnt

The failure of the joint venture between GE, USA, and Triveni Turbine in the arena of steam turbines within two years of its formation clearly teaches us an extremely valuable lesson for future developments in the high technology segments. Some of the most imperative lessons are enumerated below:

- No OEM, irrespective of its originality, will provide complete 100 per cent technology for low bypass jet aeroengine of 110-130 KN thrust niche technology design, development, and prototype manufacturing at any consideration. It is important to note that the OEMs have invested a huge number of resources in the design and developmental processes. Moreover, it is a constant source of revenue for the prolonged period or entire service life of the aeroengine from all their global customers. As evident, the aeroengine is a niche technology under the control of only a few select countries (namely the USA, France, the UK, and Russia), and almost all of the civil and military aeroengines are produced by these countries.

- Some of the technologies, processes, and modules of the aeroengine are provided by a third party holding Intellectual Property Rights (IPRs), and these technologies cannot be provided to Indian enterprises by the aeroengine OEMs.

- Most OEMs use engine consumables from their associated companies or countries, and recommend the mandatory continual usage of the same products rather than consumables with generic characteristics or equivalent grades.

- All the OEMs will charge additional costs for future hardware and software upgrades, improvement of maintenance processes, documentation upgrades, inspection and defect investigation, analysis of data, lifetime support of technical spares and equipment, and obsolescence management.

- The OEMs, if they collaborate on a joint aeroengine development project, will definitively prohibit Indian companies from exporting the technology or the aeroengine as the final product to any foreign-friendly country where their direct revenue or interest is affected.

Hence, it is imperative to keep our interest and options open for the future and attempt to proceed with specific and limited technology development, or ToT, and definite or collaborative arrangements for joint aeroengine design, development, and prototype manufacturing in the specific domains, modules, and aggregates only. India should not go for complete joint or collaborative development of aeroengines since, in the long run, it will definitively be highly counterproductive. The critical technologies that we are aware that the OEMs will definitively not provide need to be developed through a consortium of Indian industries, institutes of national repute, and government laboratories.

Presently, India has a large industrial base with capabilities in advanced tool design, computer numeric controlled machines, computer-aided software simulation tools, coordinate measuring machines, design simulator validation experts, a centre for precision engineering machining facilities, and a large number of industries with 4.0 industrial standards. There are a number of Indian industrial enterprises, in the public as well as private sectors, that are already engaged in the various aeroengine parts manufacturing for the OEMs of the aeroengines. Some of these companies are also engaged in the most critical processes of aeroengine manufacturing, such as turbine disc machining and main bearing housing machining for the various OEMs. Indian companies are also excelling in manufacturing a large variety of fasteners for aviation companies such as Boeing and Airbus, beating more than 250 of the best companies from other parts of the world.

It is a well-known fact that India is a large-scale automobile spare parts and major assembly manufacturer and supplier to most reputed and established automobile manufacturers across the world. India also has a strong base for wind turbine manufacturing through major companies such as Siemens Gamesa, Suzlon, Wind World India, Regen Powertech Private Limited (ReGen), Innox Wind Limited, Orient Wind Power Limited, Indowind Energy Limited, RRB Energy Limited, Vesta India, Enercon India Private Limited, and others. It may be important to note that the global wind power energy market segment was worth US$ 62.1 billion in 2019 and is likely to rise to US$ 127.2 billion by 2027, with a compound annual growth rate of 9.3 per cent from 2020-2027.

There are many manufacturing and engineering processes such as forging, casting, machining, polishing, protective coating, vibro-tumbling, broaching, grit/sand blasting, chemical etching, electron beam welding, robotic thermal spray, heat treatment, non-destructive tests, etc. with respect to the compressor and turbine casing, turbine rotor, turbine discs, turbine stator, and rotor blades of steam and wind turbines, which are similar to the aviation turbines. Thus, it is feasible to get these processes outsourced for the aeroengine parts to these well-established companies to ensure high-quality products and also save time and cost.

Gas turbines are also the most efficiently used machines for electrical power generation, chemical plants and refineries, cement plants, paper plants, and sugar industries. Gas turbine engines are also the backbone of military and civil aircraft propulsion due to their high thrust-to-weight ratios.

India is in the process of designing and developing a fifth-generation stealth fighter aircraft known as the AMCA and is likely to begin serial production by 2035. The requirement for the jet engine is likely to be 110-130 KN thrust, and, hence, GE-404 and GE-414, presently being used on the LCA Tejas with a thrust of 85 KN and 98 KN, respectively, will not be suitable. Perforce, India has to design and develop 110-130 KN aeroengines suitable for the AMCA as early as feasible.

Based on the discussion with GTRE specialists recently, it appears that GTRE is not in a position to design, develop, and undertake prototype manufacturing of 110-130 KN modular core aeroengines indigenously primarily due to inadequate technical knowhow and, thus, requires either a joint venture, co-development, or design ToT from an established foreign based original equipment manufacturer on priority. The inability to develop a suitable modular core turbofan engine of 110-130 KN indigenously will have an extremely adverse impact on the development of the AMCA since it will result in our continued dependence on the foreign-based OEM for the entire service life span of the aeroengine if procured from abroad for smooth engine operation, spare parts, consumables, upgrades, service life revisions, defect investigations, reliability issues, and even operational restrictions. In addition, it will result in a huge financial burden on the country in terms of initial procurement and subsequent sustainability issues. Moreover, in the event of non-conformance on any major issue with the OEM's country, whether political, diplomatic, or realignment of the new world order, India can be starved of technology upgrades, spare parts, and aeroengine consumables, and our air operations can be adversely affected. Hence, it is imperative to fast-track our indigenous aeroengine design and development efforts in a mission mode involving India's top industrial companies in the private and public sectors, academia from national and international institutions of repute, and experienced engineers from various spheres with rich expertise. The control of the mission mode developmental agency should be given to outstanding engineers with a proven track record, and the development must progress in a chronological order with a strict time-bound schedule, without any administrative or bureaucratic hurdles.

6

Aeroengine Development at GTRE

The Gas Turbine Research Centre (GTRC) was created at 4 Base Repair Depot, Air Force Station, Kanpur, in 1959. It was brought under the Defence Research and Development Organisation (DRDO) in 1961 and shifted to Bangalore, and the name was changed to Gas Turbine Research Establishment (GTRE). GTRE is actively involved in the design, development, and prototype manufacturing of gas turbine engines for military combat aircraft. It is understood that GTRE was instrumental in the design and development of a centrifugal compressor type 10 KN thrust engine between 1959 and 1961.

GTRE subsequently undertook the design and development of a 1700K reheat system for the Orpheus 703 engine to enhance its power, which was duly certified in 1973. GTRE was also engaged in the dry thrust improvement programme of the Orpheus 703 engine by replacing the subsonic compressor with a transonic compressor.

GTRE commenced the design and development of a demonstrator gas turbine engine named GTX37-14U for fighter aircraft, and the performance test was carried out in 1977. It is understood that the demonstration phase was completed in 1981, and this engine was optimised to build a new low Bypass Ratio (BPR) jet engine for multirole combat aircraft code-named GTX 37-14U B.

GTRE has adequate facilities in terms of laboratories, engine test-beds, measuring instruments, sensors, and other infrastructure for aeroengine design, development, and prototype manufacturing. Some of the prominent specialised equipment is mentioned as follows:

- Full-scale compressor test facility.
- Full-scale combustion chamber test facility.
- Full-scale turbine test facility.
- Flow visualisation facility.
- Afterburner combustion test facility.
- Heat transfer studies test facility.
- High mass flow high pressure air supply facility.
- Normally aspirated engine test facilities (four test cells).
- High Mach number engine test facility.
  - Bird ingestion (static and dynamic) test facility.
- Foreign object containment test facility.
  - Engine casing test facility (axial, torsional and radial loads).
- Ferris wheel test facility.
- Engine gearbox back-to-back facility.
- Thermal fatigue test facility.
- Universal Testing Machine (UTM) test facility.
- Vibration shaker with sine excitation facility.
- Aircraft cockpit environment control test facility.

Pursuant to the Government of India's initiation for the development of the indigenous LCA, a committee of experts from the ADA, HAL, and GTRE decided to make certain changes in the aeroengine GTX37-14U already under development at GTRE for usage on the LCA aircraft. This led to the design and development of a new turbofan engine code-named 'Kaveri', suitable for installation on the LCA. The Government of India sanctioned the Kaveri engine design and development programme and allotted an amount of ₹392.61 crore, with an initial project completion date of December 1996. Since aeroengine design and development is a complex process involving a high degree of technological challenges, including the requirement of advanced materials, state-of-the-art sensors, and precision engineering processes, the project cost and time-frame were revised several times.

The Kaveri Aeroengine Development Programme was sanctioned by the Government of India in March 1989 with a project cost of ₹382.81 crore at the 1987 price level. The probable date of completion was given as December 1996. The probable date of completion and the cost of the project were revised several times up to December 2009 and ₹2,839 crore, respectively, with ₹2,105 crore [Foreign Exchange (FE) element of ₹1,227 crore] earmarked for interim flight trials with the LCA Tejas and the balance ₹734 crore for the development of the final flight standard engine. The following developmental work has been done on the Kaveri aeroengine programme:

- Nine full prototype aeroengines and four core engines have been built.
- A total of 2,882 hours of engine testing have been completed.
- Kaveri aeroengines have been tested for 73 hours of altitude testing at the Central Institute of Aviation Motors (CIAM), Russia, and 57 hours of flight testing on IL-76 aircraft at the Gromov Flight Research Institute (GFRI), Russia.
- The engine has attained technical maturity, as concluded from the Kaveri engine technical audit by M/S Safran Aircraft Engine Ltd.
- Twelve materials, including titanium, steel, and superalloys, have been indigenously developed.
- The engine is likely to be used for unmanned aircraft.

As part of the Kaveri engine development, GTRE designed, developed, and manufactured nine full-fledged engines and four core engines, subjected them to the ground test-bed run, and sent one of the aeroengines to Russia for the flight test-bed performance test. However, the aeroengine did not pass all the specified tests. It is learnt that probably the aeroengine did not produce the desired dry and wet thrust at the rated RPM and had other related problems such as aeroengine overweight and excessive engine vibration.

Even after over three decades on the Kaveri aeroengine programme, GTRE did not succeed in producing a successful engine suitable for installation on the LCA Tejas, thus, forcing the ADA to go for an alternate engine from GE, USA, and it selected the GE-404-IN-20. It is understood that the LCA II and the AMCA are likely to be powered by a more powerful engine, the GE-414, with a thrust of 98 KN.

Import of Critical Components

As per the Ministry of Defence clarification, "a few of the accessories like the lubrication pump, integrated nozzle actuation system and variable guide vanes actuation system, etc. are being imported. The expenditure on imports is to the tune of ₹500 crore (approximately), which is 25 per cent of the total sanctioned project cost of ₹2,105 crore. However, the import content will be further reduced once the Kaveri engine is under production." The Public Accounts Committee (PAC) report remains silent on the exact reasons for the failure of GTRE efforts toward the design, development, and manufacturing of the Kaveri aeroengine. There is no denial of the fact that the design, development, and prototype manufacturing of a low bypass turbofan aeroengine suitable for installation on a high-performance fighter aircraft is a complex technological challenge. Some of the prominent reasons outlined by the top management of GTRE that are causing hindrances in aeroengine development in India are mentioned below:

(a)   High technical complexities of a turbofan engine.
(b)   Non-availability of requisite aerospace-grade material for aeroengine manufacturing in India resulted in a long lead time involved in importing the raw materials from abroad.
(c)   Non-availability of polymer material for making rubberised seals, gaskets, etc., for withstanding high temperatures.
(d)   Non-availability of altitude and flight test-beds in India and, thus, sending the aeroengine abroad for testing, resulting in delays.
(e)   Non-availability of precision engineering facilities for machining the compressor and turbine blades.

Although the following aspects are not covered in the PAC report, it is learnt through unconfirmed sources that the Kaveri aeroengine failed primarily on the following accounts:

(a) The aeroengine did not produce the specified dry thrust and wet thrust as envisaged.

(b) The weight of the aeroengine was more than the specified limit. The weight of the Kaveri aeroengine for the first prototype was approximately 1,433.7 kg. GTRE attempted to introduce carbon fibre material, and the weight of the aeroengine was reduced to 1,235 kg. However, the weight of the Kaveri aeroengine was still far beyond the acceptable limit for the LCA Tejas aircraft requirement, i.e., 1,100 kg.

(c) Inadequate technical expertise at GTRE.

(d) Inadequate performance of the Kaveri aeroengine at high altitudes.

(e) Combustion flicker.

(f) Insufficient thrust (both dry thrust and wet thrust) and screech noise at the ground, as well as high altitude testing from the afterburner.

(g) The flow of funds for the Kaveri aeroengine development did not match the developmental cycle.

(h) The thrust-to-weight ratio of the Kaveri aeroengine seems to be much lower than that of other contemporary aeroengines in the world. It has been determined that the Kaveri aeroengine thrust-to-weight ratio is approximately 5.7. In contrast, the M-53 P2 aeroengine installed on the Mirage-2000 aircraft, which was designed and developed in 1973, has a thrust-to-weight ratio of 6.4. The M-88 aeroengine powering the latest generation fighter aircraft Rafale has a thrust-to-weight ratio of approximately 8.5.

(i) The most probable reason seems to be a higher BPR in the M-53 P2 and M-88 aeroengines in comparison to the Kaveri aeroengine. However, the exact dates of the two aeroengines (Kaveri and M-88) are not available to substantiate the arguments at this stage. Although the open-domain information states that the Kaveri aeroengine BPR is 0.16:1, for the M88 aeroengine, it is 0.3:1. As a common practice, most low bypass aeroengines have a BPR ranging from 0.2:1 to 0.5:1. It is important to note that the EJ-200 aeroengine has a BPR of 0.4:1, and the BPR of the GE-404 powering the light combat aircraft, Tejas, is 0.34:1. It seems the engine core size selection by GTRE for the Kaveri aeroengine was probably larger than what was actually required. This aspect might have caused the higher weight of the aeroengine and a lower thrust-to-weight ratio.

It is important to note that while a few reviews were done by IAF officers pertaining to the progress of the Kaveri aeroengine at GTRE, only a single audit of the account and project progress update was done by the Controller of Defence Accounts and the PAC during the entire project cycle of over 33 years of the developmental phase of the Kaveri aeroengine. There seems to be a clear absence of strong leadership, high degree of commitment, clarity of accountability, resolute determination, the necessary clear and concise direction, supervisory checks, a lack of incentive or enthusiasm, a lack of consultation or discussion on professional matters with specialists with domain knowledge and academicians, and a lack of project monitoring and control mechanisms, which are responsible for the failure of the Kaveri developmental engine project.

It may be noted with great concern that GTRE could not develop a successful aeroengine in more than three decades of the developmental phase of the Kaveri aeroengine due to various reasons that may or may not be attributable solely to GTRE directly. However, since GTRE has incurred a huge expenditure to the tune of over ₹2,100 crore of the taxpayers' money, the country could have derived some spinoff benefits in terms of technology development for marine engines or ground-based turbine technologies for industrial applications, including power plants, smaller jet engines for small commercial transport aircraft, or converting the turbofan engine to a turboprop or turboshaft for wide-range applications for the Indian aviation segment. GTRE could have also immensely contributed to the improvement of the ageing and obsolete aeroengines of the IAF combat fighter aircraft, such as the R-25 aeroengine of the MiG-21 or the R-29 aeroengine of the MiG-23/27, in terms of the integration of state-of-the-art sensors for better health monitoring related parameters, resulting in the probable avoidance of servicing, maintenance, and obsolescence-related incidents or accidents.

It is highlighted that on MiG-21 aircraft, the oil bottle is located behind a few rigid steel pipelines, and in order to check the engine oil level, only a technician with a thin fist can cut the locking wire, remove the dipstick, and check the engine oil level. GTRE probably could have facilitated easy access to oil bottle inspection procedures either by relocating the oil bottle or rerouting the steel pipelines.

The scientists and engineers from GTRE individually display exemplary passion with a high degree of professionalism. However, while working as a team, zeal and enthusiasm need to be maintained at either a higher than the desired level or at least in line with the expected level. It is important to note that while working as an individual, the person gets rewarded for his notable deliverables. In contrast, in the case of teamwork, either the entire team or the team leader is recognised. Thus, the performance of the individual sometimes varies when he is working on a project individually or as a team member. DRDO, like any other government organisation, provides adequate opportunities to its employees for individual growth and career progression, wherein many scientists pursue Master's and PhD degrees from the most reputed Indian institutes. However, in the pursuit of career progression, it has been observed that some of the individuals produce several papers that lack quality and do not find a place in internationally reputed journals. It has also been observed that in a few cases, the same papers have been presented at different forums, with minimal changes incorporated for the sake of the number of papers presented, which may help individuals' promotability. The much-needed free flow of pure technological discussion with practising professionals and academicians in the garb of a high degree of secrecy and resorting only to in-house closed-door meetings have contributed towards hindrance in early resolution of technological problems, resulting in unprecedented delays, project cost overruns, and, finally, the closure of the Kaveri aeroengine developmental project, without deriving any direct benefit to the country.

7

Incorporation of Advanced Technologies

The quest for the ideal and more efficient aeroengine and continuous improvement pertaining to aerodynamic, thermal efficiency, fuel saving, higher reliability, and environmental conformance through pollution and noise reduction is being progressed by all the designers and aeroengine manufacturers. Most aeroengine manufacturing companies and leading academic institutions are working towards carbon-free aviation aeroengines by developing disruptive technologies that may result in ultra-efficient aeroengines for the future generation of aircraft propulsion.

The advanced technologies in the area of new generation aeroengines are likely to concentrate on the following domains predominantly:

(a)   Aeroengines with carbon-free exhaust gases.

(b)   Higher thrust-to-weight ratio.

(c)   Higher compression ratio.

(d)   Reduction of noise pollution.

(e)   Search for alternate sustainable fuels such as hydrogen, methanol, or solar.

(f)   Nacelles design for better aerodynamics.

(g)   Active control compressor and turbine tip clearance technology.

(h)   Higher use of advanced composite material for reduction of aeroengine weight.

(i)   More efficient Full Authority Digital (Electronic) Engine Control (FADEC).

(j)   Reduction of mechanical power transmission losses.

(k)   Advanced metal alloys to withstand higher turbine inlet temperatures.

(l)   Maintenance free aeroengine.

(m) Higher safety and reliability factor.

(n)   Better coolant for faster heat dissipation.

(o)   Incorporation of stealth technology for military aeroengines.

(p)   Aeroengine Steering Revolution System (ASRS) as a replacement for canards and Aerodynamically Only Obsolete Flight (AOOF) control.

(q)   Reduction of aeroengine infra-red signature.

(r)   Reduction of aeroengine weight and size by reducing the core of the engine.

(s)   Use of Composite Carbon Matrix Composites (CAMC) for fan blades and other parts for weight reduction.

The advanced countries invest continuously in technology improvement programmes in the various aspects of the aeroengine modules that are not directly related to any specific engine development. One such example is the National Aeronautical and Space Agency's (NASA's) Glen Research Centre, which has recently collaborated with P&W for the development of a next-generation combustion chamber known as the Hybrid Efficient Core (HYTEC) project to obtain cleaner, more efficient, and more sustainable commercial flights in the future. The project cost is US$ 13.1 million, and the new combustion chamber will be subjected to testing with sustainable aviation fuel and primarily be used for single-aisle aircraft aeroengines. It is pertinent to note that the switchover from the present aviation turbine fuel to sustainable fuel is a low-hanging fruit since it will reduce the operating cost drastically for the airline's operators in addition to the reduction of pollutant gases from the aeroengine's exhaust. The United States Air Force (USAF) started an adaptive cycle or variable cycle aeroengine development programme known as the Adaptive Versatile Engine Technology (ADVENT) for next-generation military aircraft in the 2000 lbs (89 KN) thrust class in April 2007. The programme was replaced with the Adaptive Engine Technology Demonstrator (AETD) in 2012, and, finally, replaced by the Adaptive Engine Transition Programme (AETP) in 2016, with a focus on developing and testing 45,000 lbs (200 KN) thrust-class aeroengines for sixth-generation fighter aircraft, including the reengineering of the F-35 and also for the next generation of air dominance aircraft for the US Navy.

It is important to note that the F-35 is a relatively new aircraft in the USAF inventory, flying with the specifically designed and manufactured F-135 aeroengine by P&W; however, as part of an alternate aeroengine development programme, Rolls-Royce has partnered with GE for the development of a new F-136 aeroengine as part of a future engine upgrade on the F-35. The rising cost of fuel and labour is increasing the operating costs of the world's Air Forces and commercial airline operators, and, hence, the government agencies and the airline industry are shifting their focus to the use of alternate sustainable fuel, and the reduction of fuel consumption, noise levels, and emission levels. The latest research in the field of propulsion is also exploring the feasibility of electric and hybrid propulsion systems. The resultant advantages of the latest research and new simulation technologies will definitely reduce the aeroengine design and developmental cycle, engine weight, aeroengine emission and noise factors, maintenance activities, and aeroengine fuel consumption, and enhance inspection periodicity and reliability, and increase the component's service life.

Improvement of Propulsion Technology Approach

In the last three decades, aeroengine designers and engineers have concentrated more on the following points:
(a)   Increased turbine inlet temperatures.
(b)   Increased compressor pressure ratio.
(c)   Higher Bypass Ratio (BPR).
(d)   Increased fan and nacelle performance.
(e)   Reduction of noise and pollutant gases emission.
(f)   Improvement of reliability.
(g)   Improvement of maintainability.

Presently, the aeroengine researchers and engineers are working on further improvement of the above-mentioned aspects and the new aspects as described below:

(h) Advanced material and material processing techniques.

(i) Advancement in turbo-machine technology.

(j) Improvement in combustion technology.

(k) The utilisation of Computational Fluid Dynamics (CFD) in aeroengine design.

(l) Advanced sensors, including the Micro-Electronic Mechanical Systems (MEMS).

(m) Usage of magnetic bearings.

(n) Additive manufacturing for cost and time reduction.

(o) Simplifying maintenance activities.

(p) Fuel savings.

Concept of Green Engines

Environmental protection through the reduction of engine emissions and noise through technological innovation is the need of the hour. Recently, Boeing 737 and Airbus 320 variants demonstrated that the new model of the same aircraft can carry more passengers with 23 per cent less fuel. The emission of CO_2, H_2O, O_2, and N_2 is a function of engine fuel-burning efficiency in the combustion chamber of the aeroengine. The following areas may be at the forefront:

(a) Lightweight LP system for turbofans.

(b) Use of composite fan blades.

(c) High-efficiency LP spool technology.

(d) High-speed turbine design.

(e) High-efficiency and lightweight compressors and turbines.

(f) Low-emission combustion chamber.

Some of the important future design considerations can be summarised below:

(a) Ultrahigh bypass turbofan engine.

(b) Open engine rotors.

(c)   Use of sustainable alternate fuels.
(d)   Relocating engines on the body of the aircraft to deflect the exhaust upwards.
(e)   Blended wing body.
(f)   Advanced electrical power technologies.
(g)   Electrical engines using lithium polymer batteries and solar-powered manned aircraft designed to fly both day and night.
(h)   Solar electric propulsion.
(i)   Ionic electrical thrusters.

Heat Recovery Concepts

A number of independent researchers are also working on the following technologies

(a)   Combined Brayton cycle aeroengine.
(b)   Multi-fuel hybrid engine.

Presently, more than 50 per cent of the energy gets wasted in terms of heat. It is presumed that a heat exchanger integrated into the core of a turbofan engine can convert recovered heat into useful power, which can be used for onboard systems or to power an electrically driven fan to produce auxiliary thrust. A dual combustion chamber wherein the high temperature generated in the first combustion chamber allows ignition-less combustion in the second combustion chamber, thus, reducing CO and NO_x emissions. Cryogenic bleed air cooling can enhance the engine's thermodynamic efficiency.

Advanced Materials for Aeroengines

The basic design and construction of the jet engine designed by Dr. Ernst Heinkel in 1939 have not changed to date; however, tremendous improvement has taken place in terms of thrust, reliability, and efficiency of the modern aeroengine. The improvement in the aeroengine performance, specially with respect to the new material and the processing technologies for the turbine blade and the turbine disc are unprecedented. As we are aware, the turbine blades are attached to the turbine disc periphery and extend across the entire flow passage of the turbine segment. The turbine blades are designed as per the aerofoil shape. When the high velocity hot gases pass over the turbine blades, they generate lift which gets transmitted to the turbine shaft through the turbine disc which, in turn, rotates the compressor. The continual improvement in the individual major parts of the aeroengine has been more focussed on by the OEMs to attain better aerodynamic and propulsion efficiency, with higher safety and reliability. Another important factor that has caught the attention of the designers and engine manufacturers is specific fuel consumption. Aerodynamic design technology for a compressor and a turbine, cooling technology for turbine blades, and advances in material technology have significantly improved jet engine performance. With the invention of improved materials with reduced weight and better structural strength, i.e., lower density, higher strength, better workability, and heat-resistant properties, these are specifically selected for application on the aeroengine. The different materials generally used for different modules of aeroengines are covered in the succeeding paragraphs.

Fan Compressor

This is the first stage of the engine from the front and is responsible for the suction of ambient air into the engine. The earlier generation of aeroengines used to consume the entire air in the core of the engine for combustion and a part of the compressed air for cooling the combustion chamber casing and the turbine blades. The present generation of fighter combat aircraft aeroengines are generally low bypass turbofans wherein the bypass ratios are generally between 0.2:1 and 0.9:1 and the core engine provides more thrust than the bypass thrust. The fan blades are the longest and widest in order to induct the highest amount of atmospheric air into the engine, and they are made from light and high-specific-strength material. The older generation of Russian jet aeroengines used stainless steel blades for higher rigidity, which were heavier and led to higher stress on the fan disc due to high centrifugal forces. Subsequently, Rolls-Royce and Safran (Snecma) switched over

Fig 7.1: GE-90 Fan Compressor

to titanium for the fan blade due to its lightweight and better strength; however, it had a tendency to bend. Thereby, it needed a support point in the middle section of the blade. The support point is built up by welding a harder and corrosion-free material through Tungsten Inert Gas (TIG) welding so that the mating of the adjacent blades is smoother and no physical damage is caused to the blades. The latest trend in the new generation of aeroengines is to use composite fan blades comprising Carbon Fibre Reinforced Plastic Composite (CFRP); however, the leading edge of these blades needs to be strengthened by a harder material, preferably titanium (Ti-6Al-4V). This arrangement has been incorporated into the world's most powerful GE-90 aeroengine. This aspect (strengthening of the leading edge) will provide resistance from impact damage, such as foreign object damage or a bird hit.

Fan Case

The fan case is circular in nature, and one of the heaviest parts of the engine, and is generally made from aluminium-magnesium cascaded alloys or titanium. The latest trend is to use CFRP material for its lightweight, better strength, high durability, better fatigue resistance, and corrosion-free characteristics.

Compressor Module

The compressor module compresses the air inducted into the engine by the preceding fan compressor. The air compression takes place in various stages of the fan, LP and HP compressors, and the temperature also rises as the air is compressed in the succeeding stages. Thus, the material should be able to withstand high temperatures and work efficiently. Generally, stainless steel or titanium is used in the LP compressor since the temperature is relatively lower; however, normally, in the HP compressor, either a steel-nickel-based superalloy such as Hastelloy X or a Ti-6Al-4V alloy is used to withstand the higher temperature. The air temperature in the HP compressor in the larger jet engine segment can be in the region of 400-500ºC. It is pertinent to note that Ti-6Al-4V is the most appropriate alloy used for the aerospace segment and almost 50 per cent of the total titanium used is in this alloy form. It constitutes a lamellar structure in terms of mechanical properties, and, thus, it provides more resistance against creep and fatigue growth and higher toughness, and it also gives high-temperature strength below 800ºC. Titanium-based alloys such as Ti-6242 (Ti-6Al-2Sn-4Zr-2Mo) are also being used for HP compressors, mainly to reduce the weight of the aeroengine while also withstanding the high temperatures for prolonged operations.

Fig 7.2: Axial Compressor Module

Combustion Chamber

After the compressed air comes out of the HP compressor, it enters the combustion chamber, where the atomised fuel is mixed with the air and ignited to raise the temperature of the air up to 1000°C. The material used for the combustion chamber is either nickel or cobalt-based superalloys, which are heat-resistant materials. A number of additives are used to improve the mechanical properties of the superalloy,

Fig 7.3: Annular Combustion Chamber

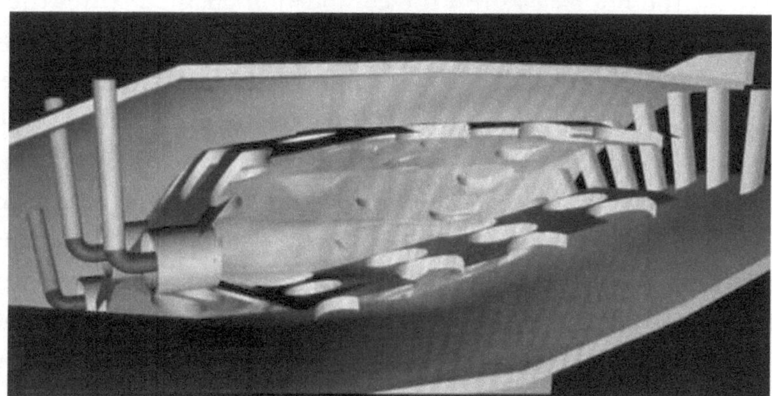

such as aluminium (Al) and titanium (Ti) for strength, chromium (Cr) for corrosion resistance, and molybdenum (Mo), tungsten (W), and rhenium (Re) for increasing high-temperature strength. Some of the prominent materials used for the combustion chamber are Hastelloy X, Nimonic 263, HA188, 617, and 230, as per the following chemical compositions:

Table 7.1

Grade	Chemical Composition	Remarks
Hastelloy X	Ni22Cr1.5Co19Fe0.7W9Mo0.07C0.005B	Ni-based superalloy
Nimonic 263	Ni20Cr20Co0.4Mo2.1Ti0.4Al0.06C	Ni-based superalloy
HA188	Co22Cr22Ni1.5Fe14W0.05C0.01B	Cobalt-based superalloy
617	54Ni22Cr12.5Co8.5Mo1.2Al	Ni-based superalloy
230	55Ni22Cr5Co3Fe14W2Mo0.35Al0.10C0.015B	Ni-based superalloy. Value for Co, Fe and B are upper limits.

Hastelloy X is a highly versatile superalloy material used for the combustion chamber, afterburner rings, flame holders, spray bars, transition duct manufacturing, other aeroengine and aircraft parts, industrial furnaces, and chemical processing applications.

Turbine Modules

The turbine module comprises a turbine disc, a turbine shaft, and turbine blades mounted on the periphery of the turbine disc as shown in Fig 7.4. The hot gases from the combustion chamber enter the HP turbine section wherein the power from the low-volume, high-pressure and high-velocity hot gases is extracted by the HP turbine blades of smaller sizes and the gases further move to the intermediate turbine (if available), and then to the LP turbine where the energy from the relatively higher volume lower pressure and lesser velocity gases is extracted by the intermediate turbine stage (if available). Since the intermediate and LP turbine blades are longer and wider, they have a tendency to bend and collide with the adjacent blades, thereby causing fatigue, stress, and vibration. Some of the designers provide additional

Fig 7.4: Turbine Disc and Blades

support at the tip of the blade in the form of a crown called a shroud, which, when joined with other LP turbine blades, forms a circular ring that helps in the reduction of gas leakage, provides support to each other, and also helps in dampening the vibrations of the blades. However, some designers feel the extra weight on the tip of the blade can create additional centrifugal loading. Some designers prefer to design LP blades with mid-term support, wherein the ring is formed in the mid-segment of the blade; however, this may affect the aerodynamic efficiency of the aeroengine.

It is important to note that the life of the turbine blades is specified as the number of cycles the blades are subjected to the maximum possible operating temperature (peak performance) rather than the total time span they operate in the stabilised conditions. It is indeed a departure from the old practice of limiting the engine operation based on total running hours. The life of the turbine blades or the hours of their operation in stabilised conditions does not impact the blade life with any severity.

Turbine Blades

The turbine module is generally exposed to the harshest environment in view of the high temperature, high pressure, and high-velocity gases

Fig 7.5: Turbine Blades

passing through it. Hence, the material required for the construction of the turbine module should be able to withstand high temperatures and work efficiently with low creep over a prolonged period of operation, i.e., high creep strength, high temperature corrosion and fatigue resistance. Generally, cobalt and nickel-based superalloys are used for manufacturing turbine blades. The turbine blades' operating temperatures exceed 1200°C, and can go up to 1400°C. Generally, the turbine blades are coated with thermal barrier coatings such as Serme-T, Sermeloy-J, and Sermetal-W, or yttria, ytterbia, or zirconia (ZrO_2 coating). In addition, the high pressure turbine blades are hollow physically through a passage from the root and have fine holes in the trailing edge through which cool air is passed to continuously cool the blades.

Turbine Shaft

The turbine module (disc with blades) and the compressor module (disc with blades) are mounted on the same shaft in the single-spool engine. In multi-spool aeroengines, the turbine and compressor are mounted on their respective shafts. All the shafts pass through the centre of the combustion chamber and are generally obscured from the direct heat; however, the transmission of heat through the process

Fig 7.6: Turbine Shaft

of conduction from the blade to the disc and to the shaft cannot be completely eliminated. The material used for the construction of the turbine shaft should have the following characteristics:

(a)   Heat resistance.
(b)   High temperature strength.
(c)   High fatigue strength.
(d)   Toughness.
(e)   Creep resistant.
(f)   High torsional strength.

Generally, steel with high purity and without any segregation is used to ensure higher reliability, such as stainless steel; however, Ti-6Al4V alloys and ferrous and nickel-based superalloys are also used in smaller engines. The most common materials used are Cr-Mo-V steels, Inco-718, and maraging steel (GE1014). The maraging steel is a very strong steel containing 18-25 per cent nickel and has high fatigue strength (approximately 22000 MPa or higher) and high toughness due to its high cleanliness. The maraging steel has been used in some of the most successful modern aeroengines, such as the Rolls-Royce Trent 1000, GE-90-115B, and GEnx aeroengines.

Future Trend

The most advanced commercial aeroengine material composition is depicted in percentile terms as follows:

(a)   Nickel (Ni) based alloys   47 per cent
(b)   Titanium (Ti) alloy   25 per cent
(c)   Iron-based alloy   16 per cent
(d)   Aluminium alloy   08 per cent
(e)   Composite   04 per cent

In the futuristic aeroengine, it is expected that the use of lighter materials such as aluminium, titanium, and composites will increase substantially. Aeroengines in the present configurations are likely to continue for a longer time-frame until they are possibly replaced by energy-efficient, low-emission alternatives to conventional propulsion systems in the form of ionic wind thrusters. The ionic wind thrusters are likely to use ionic energy, which is created when a current passes between two electrodes. However, the ionic wind thrusters are still in the basic research stage, and, hence, technology maturity and commercial exploitation cannot be estimated or predicted at this stage.

8

Ground Testing and Air Flight Testing

The aeroengine is analogous to the human heart for the aircraft, and aircraft safety is of paramount importance since it is directly related to human safety. The performance of the pure jet and turbofan aeroengines is measured in terms of thrust in KN, kg, or lbs, which is produced by the engine at the jet nozzle. The higher the exit gas velocity, the higher the thrust. However, for the turboprop and turboshaft aeroengines, the power is measured in terms of Shaft Horse Power (SHP). The thrust, as well as the SHP of the aeroengine achieved, are for a given weight, fuel consumption, and the diameter of the air inlet of the aeroengine. The thrust or the SHP of an aeroengine is directly dependent on the mass of the airflow into the aeroengine, and it is important to note that the performance of the aeroengine is also dependent on the ambient pressure, temperature, and altitude of the test-bed.

In order to ensure the most economical fuel consumption, thereby resulting in cost savings for operations of the aeroengine, the engine performance is also calculated in terms of Specific Fuel Consumption (SFC). SFC is the ratio of fuel consumption to thrust or SHP. The SFC unit is measured by calculating the kg of fuel per hour per kg of thrust, or SHP. The lower the numerical value of SFC, the better the performance of the aeroengine. The size of the aeroengine core largely depends on thermal (internal) and propulsive (external) efficiency. In order to make a direct comparison of different aeroengines, all the parameters are converted to the International Standard Atmosphere (ISA), which is based on the temperature drop rate of approximately 1.98 K (kelvin) per 1,000 ft drop in altitude. The aeroengine designed and manufactured in India must be able to perform throughout the flight envelope of the aircraft at the designed thrust and the specified RPM. Post-manufacturing, the aeroengine is subjected to stringent tests, and the performance must be monitored for certification by the relevant competent authority. The Federal Aviation Administration (FAA) Part 14 Code of Federal Regulations is applicable for engine and propeller testing, along with Advisory Circulars Part 33 (Airworthiness Standards: Aircraft Engines), Part 34 (Fuel Venting and Exhaust Emission for the turbine engine powered aeroplane). The test protocol is designed, and all the tests on the ground engine test-bed are performed in the presence of certification authorities. The main parameters that are crucial for the aeroengine certification are checked and recorded for all the throttle settings based on which the aeroengine performance is evaluated.

The main parameters to be observed and recorded as standard parameters are mentioned below:

- P1 ambient pressure at the air inlet of the engine.
- T1 ambient temperature at the air inlet of the engine.
- P2 air pressure after the fan compressor or at the entry for the LP compressor.
- T2 air temperature after the fan compressor or at the entry of the LP compressor.
- P3 air pressure after the LP compressor.
- T3 air temperature after the LP compressor.
- P4 air pressure after the HP compressor.
- T4 air temperature after the HP compressor.
- P5 combustion chamber pressure.
- T5 combustion chamber temperature.
- P6 HP turbine inlet pressure.
- T6 HP turbine inlet temperature.

- P7 LP turbine inlet pressure.
- T7 LP turbine inlet temperature.
- P8 exhaust gas pressure.
- T8 exhaust gas temperature.
- N1 LP rotor RPM.
- N2 HP rotor RPM:
  - Air flow rates.
  - Fuel flow rates.
  - Main oil pressure.
  - Main oil temperature.
  - Scavenge oil pressure.
  - Scavenge oil temperature.
  - Engine vibration.

Based on the measurement of various parameters as mentioned above on the ground engine test-bed and flight test-bed at various test altitudes at specified aircraft speeds, the performance markers of the engine are calculated, such as SFC in various modes or ratings of the engine, aeroengine BPR, aeroengine overall pressure ratio, thrust to weight ratio, compressor, combustion chamber, turbine, and jet nozzle efficiency, etc. The numerous parameters pertaining to the type of oil used for lubrication of turbine engines, such as viscosity, viscosity index, flash point, pour point, demulsification, copper corrosion, acid number, foam characteristics, oxidation characteristics, water separation, rust protection, cleanliness, and water contents, are checked as per the specification and ASTM D-943.

The oil system in a jet turbine engine works on the dry sump principle and serves the following purposes:

- The oil system provides lubrication to the bearings through 3-4 oil jets placed at 120-90 degree circular separation on the compressor-turbine rotor shaft of 1-4 mm jet orifice diameters, depending upon the quantity of oil required for lubrication. It provides a 0.1-0.5 micron film on the rotor and ensures the rotor floats on this film without metal-to-metal direct contact.

- The oil system also takes away the heat generated at the support bearings and ensures the proper temperature of the bearing's elements (such as balls, rollers, cages, etc.). The bearing seals are designed and installed in such a way that the hot air should not come into direct contact with the lubricating oil. Although a small amount of hot air tapped from the compressor stage is also used for sealing the lubricating oil leakage at the labyrinth, which may come in contact with the lubricating oil, the quantity of the hot air should be minimal; otherwise, the engine will have oil darkening problems frequently. The quantity of hot air is controlled through a seam of a specific size installed during manufacturing or during the Repair and Overhaul (ROH) of the engine. Generally, at the labyrinth, the compressed air pressure is kept higher than the oil pressure by design, and, thus, it prevents the oil from leaking through; however, a very small amount of hot air comes in contact with the oil, which is separated in the de-aerator located on the oil tank. The scavenge oil temperature is higher than the main oil pressure, and the same is cooled in the fuel-oil heat exchanger, wherein the lubricating oil is cooled while the fuel gets heated up. The fuel-oil heat exchanger consists of narrow metallic tubes through which the pressurised fuel passes, and the hot oil flows on the external surface of the metal tubes; thus, while the oil gets cooled, the fuel gets heated up.

- The scavenge oil system takes away the minute metal debris caused by normal wear and tear and, thus, cleans the bearings and the rotor for longer, trouble-free operations. In case the metal debris is excessive, the same is to be analysed for the type of possible metal, and the engine oil needs to be flushed out and replaced.

- The oil scavenging system also has a magnetic metal detector, which alerts the pilot through a light indicator showing the presence of ferrous material in the oil and, thus, prevents further damage to the engine. As the aeroengines work on the dry sump principle, the oil scavenge rate is generally 3-4 times higher than the pressure oil.

The general calculation of the aeroengine thrust can be used by using the following formula:

$$\text{Thrust} = \text{Area (A)} \times \text{Pressure (P)} + M\,\frac{Vf}{g}$$

Area (A)   = Cross-sectional area of the point on the engine duct where the airflow is being measured in cm^2

Pressure (P) = Cross-sectional area of the specific point on the engine duct where the airflow pressure is being measured in kg/cm^2

M   = Mass flow in kg/sec

Vf   = Velocity of flow in m/sec

g   = Gravitational force in $9.8\,m/sec^2$

The formula for the thrust calculation can be used to calculate the thrust at various points or sections of the aeroengine in order to calculate the efficiency of the compressor, combustion chamber, turbine, and jet nozzle.

Afterburner Performance

The military aeroengines are accommodated inside the central fuselage of the combat aircraft for better manoeuvrability. Hence, the frontal opening dimensions are limited, unlike in transport or commercial aircraft, where the aeroengines are installed on the wings. The frontal area can be much larger and is, thus, most suitable for high bypass aeroengine installations. The military aircraft can have a low bypass engine with an afterburner or reheat arrangement for thrust augmentation so as to operate from a shorter runway or even a semi-prepared runway during operations and, thus, are provided with afterburner fitment inside the jet pipe between the LP turbine and the exhaust jet nozzle. The turbine exhaust gases mix with the bypass airflow, which is rich in oxygen, where the additional fuel is burnt in order to raise the temperature of the exhaust gas, thereby enhancing the volume of the exhaust gas at the jet nozzle and resulting in higher thrust augmentation. The temperature of the afterburner flame can be in excess of 1500°C. Thus, the walls of the jet pipe are continuously cooled by the turbine exhaust gases, which are directed to flow along the jet pipe wall and provide a thermal barrier from the reheat flame. The reheat flame is concentrated on the central axis of the jet pipe. The velocity of the turbine exhaust gases is generally between 225 and 375 m/sec, which will blow away the afterburner flame. Thus, the velocity is reduced through the diffuser, and the pressure is increased. A set of 2-3 flame stabilisers is also provided downstream to reduce the velocity of exhaust gases further in order to make it conducive for the flame to sustain and stabilise. The fuel for the afterburner is supplied from the afterburner fuel pump at a higher pressure and sprayed through the very small orifices in the main circular concentric fuel manifolds. A separate igniter provides sparks at a rate of 50-100 sparks per minute.

The enhancement of thrust augmentation is the square root of the ratio between the gas temperature after the afterburner and the exhaust temperature of the LP turbine.

$$\text{Thrust Augmentation} = \sqrt{\frac{\text{Gas temp after the afterburner}}{\text{Gas temp after the low pressure turbine}}}$$

In a modern aeroengine such as the F-135 installed on the F-35 combat fighter aircraft, the thrust augmentation achieved is approximately 50 per cent. However, it is important to note that when the afterburner is not in use, the flame stabilisers, fairings, interconnectors, etc., act as retarders, and the thrust of the aeroengine is slightly lower in comparison to a similar aeroengine without the afterburner. The performance of the afterburner system must be thoroughly checked during the ground test-bed run, as well as the air test-bed run at various altitudes in terms of thrust augmentation and specific fuel consumption. Since the fuel consumption during the reheat engagement is 2-3 times higher than the normal fuel consumption at flight idle, the reheat engagement is normally done by the pilot for a shorter duration, such as take-off and climb combat manoeuvring.

Thrust to Shaft Horse Power Calculation

As mentioned above, the thrust of the turbofan or turbojet engine is expressed as kg, KN or lbs because the thrust is a reaction force. However, the turboprop and turboshaft engines' power is expressed in Shaft Horse Power (SHP) since Power = Force × Distance. Thus, it is not feasible to compare a turboshaft or turboprop engine with a turbofan or pure jet engine directly. Hence, in order to compare the turbojet/turbofan engine with the turboshaft/turboprop engine, we need to convert the thrust (force) to work or work to thrust (force), and while doing so, we need to take the aircraft speed into account.

$$\text{Thrust Horse Power (THP)} = \frac{\text{Thrust (Kg) X Speed of aircraft}}{600}$$

(Speed of the aircraft to be written as in m/sec)

Propulsive and Thermal Efficiency

The propulsive efficiency of the jet engine is the conversion of kinetic energy to propulsive work. It is also called external efficiency, which is based on the most efficient conversion of the thermal energy (heat) to the jet velocity of the exiting gases or the kinetic energy, resulting in the most efficient way of propelling the aircraft forward as a reaction to the jet velocity. Thermal energy efficiency, also called internal efficiency, is based on the pressure ratio of the engine and the combustion chamber's maximum permissible temperature, which is restricted due to the combustion chamber material, which can only withstand up to the specified temperature. It is important to note that the diameter of the turbofan engine will be smaller and even the length will be shorter since the bypass engine is deriving the thrust from the core as well as the bypass air, and the core engine is dealing with a part of the air mass inducted into the engine, whereas for the pure jet engine since it is handling the entire amount of the air intake into the compressor, the size of the HP compressor and the HP turbine will be larger and even longer. Thus, for the same amount of power, the turbofan engine will be lighter in weight than the pure jet engine.

Factors Affecting the Aeroengine Performance

The aeroengine's performance can get affected under various environmental conditions, which are discussed in detail below:

Effect of Altitude

As we are aware, the air pressure and temperature decrease when the altitude increases since the density of the air varies.

$$\text{Density} = \frac{\text{Mass}}{\text{Volume}}$$

When the air pressure increases, the air molecules tend to move closer to each other and thereby increase their density, resulting in an increase in thrust. The air inlet of the aeroengine is constant, and, thus, at the constant speed of the engine (RPM), the mass of the air will remain constant. The density of the air affects the thrust proportionally. It is important to note that when the pressure goes up, the density goes up; however, when the temperature goes up, the density goes down, and, thus, with the temperature going down, especially at high-altitude airfields, the thrust of the aeroengine can go down by 20-25 per cent.

Effects of Aeroengine Speed

The aeroengine RPM controls the total mass flow of the air inducted into the aeroengine inlet. In order to increase the engine power (thrust), the fuel input to the engine is increased, and, thus, the aeroengine speed increases and the resultant thrust increases accordingly. At higher engine speeds, even smaller increments in the engine speed increase the thrust much higher, whereas at lower engine speeds, even if the engine speed is increased, the thrust increase is not significant.

Humidity Effects

The effect of humidity on the jet aeroengine is not significant since, generally, the amount of air inducted into the engine is much higher than what is utilised for combustion. A significant amount of air is used to cool the various parts of the engine. Thus, when the air-to-fuel ratio reduces, the cooling air compensates for the reduction of air available for combustion due to humidity.

Temperature Effects

The density of the air increases when the atmospheric temperature decreases during cold days, and thereby, the mass of air entering the engine at a given RPM will be higher, resulting in higher thrust. However, during a hot day, when the atmospheric temperature goes up, the air molecules tend to move apart, the density of air decreases, and the mass of air inducted into the engine at a given RPM reduces. Thereby, the resultant thrust will decrease accordingly. If the atmospheric temperature goes to 40°C, the thrust may decrease by 20-25 per cent.

Effect of Water Injection in the Engine

As explained earlier, the thrust of the engine reduces due to a higher atmospheric temperature. The same can be augmented by injecting water or a mixture of water and alcohol into the air intake by 10-15 per cent. When the water is injected into the aeroengine, the mass of the water and alcohol is added to the mass of the air already inducted into the engine. The water and alcohol mixture cools the mass of air inducted into the engine and maintains the air pressure by adding the water molecules to the mass flow.

Effect of Dust Accumulation

Dust accumulation is a normal phenomenon during the normal operation of the aeroengine, especially in dusty environmental operating conditions. In the desert area of operation, the dust consisting of silica particulates strikes the first stage of the compressor during air intake. Also, it causes minor erosion and roughness on the leading edge of the first-stage compressor blades over a period of time. Since the dust particles also stick on the rotor and stator blades and reduce the cross-section area of the air intake slightly, the resultant thrust also reduces.

The OEMs generally recommend compressor washing periodically to ensure proper operation of the aeroengine and the same needs to be adhered to by the operators.

Effects of Ram Air

The ram air effect is more prominent in the case of supersonic flights of high-performance military aircraft since the airflow becomes compressible at the supersonic speed of the aircraft, the mass flow increases at an increasing rate, and, thus, the thrust of the aeroengine increases at supersonic speeds. It is important to note that at higher speeds, the difference between the velocity of air at the jet nozzle and the air intake at the entrance of the aeroengine decreases (i.e., there is a decrease in acceleration). The thrust will tend to decrease as the aircraft increases its speed, however, since more mass of air is inducted into the air intake. Hence, it results in either a neutral or nominal enhancement of thrust at subsonic air speed.

Sensors Requirements

The aeroengine operation and control mechanism is based on the accurate physical sensing of the various aeroengine parameters through state-of-the-art, reliable sensors. The entire concept of an intelligent, smart, or active control aeroengine solely depends on the accuracy and repeatability of the sensor output, which is extremely important for the aeroengine control loop operation. The critical functionality of the aeroengine also governs the type of sensor to be used; for pressure, vibration, RPM, and temperature sensing, the state-of-the-art sensor must be of Technology Readiness Level (TRL) 9. In contrast, for the monitoring of parameters that are not critical, a lower TRL can be considered if the parameter is used for direct monitoring only and is not likely to be used for the engine or its components' feedback control loop systems. All the sensors, whether installed on the aeroengine, ground engine test-bed or air flight test-bed, must be able to function satisfactorily in terms of range, bandwidth, resolution, and accuracy in the entire environmental range of the aeroengine operational envelope. However, most of the sensor manufacturers adhere to the specifications of the aircraft and aeroengine as per MIL-STD 810F. Generally, the specification of most modern sensors with integrated electronics is limited to environmental conditions between -65°C to +115°C. In the distributed control architecture, the futuristic smart sensors are likely to be used for monitoring RPM, temperature, pressure, etc. and should be able to operate at very high temperatures satisfactorily well within the specified tolerances, process the acquired data, take decisions, and provide feedback for actuation of the specific system control components. Presently, all these parameters are linked and provided to the FADEC system, which receives and processes the data and provides feedback to the specific actuators in accordance with the aeroengine control law. As the technology matures for the Micro-Electronics Mechanical System (MEMS) fabricated with silicon-based technologies, various types of temperature and pressure sensors may find active roles in the distributed technologies with a faster response time-frame for system actuation. These sensors are also likely to be embedded in the various aeroengine parts for providing active predictive maintenance data and even alerting the pilot in case of a likely emergency failure of the flight-critical systems. The present sensors, which are in use on aeroengines and other industrial applications, are covered in brief below.

Temperature Sensors

Most of the high-temperature sensors used for measuring high temperatures use the thermocouple principle, wherein when two dissimilar materials are joined together, and the junction is either heated or cooled, a small voltage is generated in the electrical circuit of the thermocouple and the same is measured. The measured voltage indicates the corresponding temperature.

Piezoresistive Sensors

The piezoresistive sensors can measure the force by sensing the proportional change in the electrical resistance value of the semiconductor material, such as silicone. These sensors are generally used for measuring pressure or force.

Piezoelectric Sensors

The piezoelectric sensors are based on the generic principle, wherein piezoelectric is the property of the specific material that can generate electrical voltage when mechanical force is applied to the specific material. It is used for the measurement of acceleration, pressure, and strain in the material. The specific ceramic materials that are commonly used for piezoelectric sensors are lead zirconate titanate (PZT), barium titanate (BT), and strontium titanate (ST).

Capacitive Sensors

Capacitive proximity sensors are non-contact sensors that can detect the presence or absence of any object, irrespective of its material. The electrical property of the capacitor and the change of the capacitance value based on the electrical field around the active face of the sensor is measured. It has wide sensing applications, such as,

- Flow
- Pressure
- Liquid level
- Gap or spacing
- Thickness
- Ice detection
- Shaft angle or linear position
- Dimmer switches
- Key switches
- X-y tablet
- Accelerometers

Photoelectric Sensors

The photoelectric sensor is a device used for the determination of the presence or absence of an object by using a light transmitter, a visual or infrared light beam source, and a photoelectric receiver. It can also detect changes in the surface condition or the orientation of an object on a production line.

Inductive Sensors

Inductive sensors are based on the principle of electromagnetic field induction to detect or measure objects. When a current flows through an inductor, a magnetic field is produced, and if any moving magnetic object is in motion, it activates the current flow as per Faraday's law of induction, which is used as an output signal to detect the magnetic object.

Inertial Measurements

The inertial measurement technology is based on a mass-spring resonator scaled down to the micro level. The detection of motion is based on inductive or capacitive non-contact sensors. These are used for the detection of the rate of rotation and acceleration of the object.

Air Flight Test of Aeroengine

The aeroengine is subjected to a ground test-bed run on the ground engine test-bed to ensure its intended function and reliability factor. The various emergencies that the aeroengine may or may not encounter during its assigned service life are also checked at the engine test-bed run. Although most of the aeroengine parameters are checked during the ground test-bed, some of the important tests that are critical from a flight safety angle, such as the altitude test, vertical acceleration, vibration at high altitude, fuel burn rate, extreme flight conditions that most of the engines will never encounter in their service life, zero-G operation, sustained flight in icing conditions, etc. are to be necessarily checked on the aircraft-based engine test-bed to obtain the real data at the actual altitude flight conditions. Moreover, the ground-based test data are also validated under varying altitude conditions. All the aeroengine design and manufacturing companies have a dedicated airborne platform for testing the newly developed aeroengine. The air flight test-bed is incorporated into the multi-engine aircraft platform, with one engine station configured and instrumented with multi-sensors for sensing the various parameters of the aeroengine under test. These parameters are comprehensive and sufficient for calculating and analysing engine operating and efficiency parameters. The following OEMs own dedicated aircraft for testing and validating commercial and military aircraft aeroengines:

General Electric

The Boeing 747-100 air test-bed is 49 years old. It has successfully tested and validated 11 engines models and 39 different variants, which include the GE-90 (the most powerful engine in the world) and the GEnx for Boeing 787 Dreamliners. The aircraft operation became unsustainable due to the non-availability of spares and was, hence, phased out. GE has inducted another Boeing aircraft, the 747-400, which is being utilised for extensive testing of GE 9X aeroengines.

Pratt & Whitney (P&W)

A Boeing 747 SP is being used for air flight tests of all the aeroengines being designed and developed by P&W.

Rolls-Royce

Rolls-Royce is using a Boeing 747-200 aircraft mounted with three RB211 engines and an experimental aeroengine test-bed station specially wired for aeroengine testing. Presently, the aircraft is being used for testing of 100 per cent Sustainable Aviation Fuel (SAF) to power Trent 1000 aeroengines in experimental test flights.

Russia

Russia is using an Ilyushin Il-76 LL aircraft as a flying engine test-bed. The aircraft was also used for Indian Kaveri engine altitude testing.

China

The China Flight Test Establishment (CFTE) operates a converted Russian Il-76 MD aircraft as a flying engine test-bed. It has tested the WS-10A (Taihang) low bypass turbofan aeroengine, which is likely to power the Chinese fighter aircraft J-10 and J-11 combat aircraft.

France

The Airbus A380 is being used as an aeroengine flying test-bed at the Airbus flight test facility in Toulouse, France. The flying engine test-bed is presently being used for flight tests of CFM's cutting-edge open fan engine architecture. The joint demonstration programme is likely to use shared flight test assets at the GE Aviation Flight Test Operations Centre in Victorville, California, USA, and the Airbus Flight Facility in Toulouse, France.

India is attempting to design, develop, and manufacture a low bypass core aeroengine of 110-130 KN at GTRE, Bengaluru; however, no actionable plan has been initiated for the design, development, and creation of an indigenous flying engine test-bed to undertake testing of indigenous aeroengines without any delay or lead time in addition to cost reduction in the country itself. It is considered imperative on the part of the Government of India to sanction a project for the design, development, and modification of a multi-engine transport aircraft having completed its 50-60 per cent useful service life (in order to reduce the cost of the aircraft) and modifying it for conversion to an aeroengine flying test-bed. Adequate expertise exists in India to undertake this task on a priority basis, and this project may also be open to public and private Indian companies to expedite the developmental process and ensure completion before the indigenous low bypass core engine is due for altitude testing and validation of aeroengine parameters in an estimated time span of two to three years.

9

Potential of Aeroengine Business in India

The IAF is the fourth-largest Air Force in the world, with approximately more than 1,850 aircraft (including 630 combat fighter aircraft) in its inventory. Most of these aircraft had been procured from the erstwhile USSR, Russia, the UK, France, and the USA. These aircraft were already installed with aeroengines from their country of origin. The ToT for MRO of the aeroengines was obtained by the IAF, and the same is being undertaken by either HAL or the integral base repair depots of the IAF. It is important to note that, as per the information in the public domain, the IAF requires 42 fighter aircraft squadrons to defend the country in the eventuality of a simultaneous war with our two prominent nuclear-armed adversaries. Three MiG-21 Bison squadrons are still in service, and these are likely to be phased out by 2025. It is estimated that presently, there are only 32 combat aircraft squadrons available in the IAF, which is grossly inadequate to engage in a major conflict with two main nuclear-armed countries sharing long land borders with India. The IAF has already placed an order for 40 LCA Tejas MK-I, 73 aircraft in the Mk-IA/II and 10 trainer aircraft configurations. The initial order of 40 aircraft will be installed with GE-404-IN20 aeroengines with approximately 85 KN of thrust, whereas the 73 aircraft in the Mk-II configuration and 10 trainer aircraft will be powered by the F414-GE-INS6 engines with 98 KN of thrust. The Indian Ministry of Defence has also floated an expression of interest request for the procurement of 114 fighter aircraft for the IAF. ADA, Bengaluru, and HAL are also in the process of developing the AMCA with twin aeroengines, which are likely to be heavier than the present versions of the LCA and, thus, require more powerful aeroengines to the tune of 110-130 KN. Thus, the IAF will have at least 123 GE-404 and GE-414 aeroengines in its operational squadrons once HAL supplies all these aircraft in the next 6-7 years. The IAF has indicated procuring at least seven squadrons (126 aircraft) of the AMCA aircraft in the post-developmental phase with an aeroengine having a thrust of 110-130 KN. Thus, the IAF will have 249 (40 Tejas Mk-I, 73 Mk-IA, 10 Trainers and 126 AMCA) twin-engine indigenous fighter aircraft in its inventory. As mentioned in a previous chapter, it is the general norm that one fighter aircraft requires 2.5 to 3 aeroengines in its service life span, and, thus, a total of 375×2.5=750 aeroengines will be required for full exploitation of indigenous fighter aircraft by the IAF in the future. The rough cost of the GE-414 engine is US$ 8.23 million as per the HAL contract with GE in August 2021. Thus, if an indigenous aeroengine is not developed in a definite time-frame, India shall be procuring aeroengines for its indigenous fighter fleet worth 750×8.23=6,172.5 million, or US$ 6.17 billion at most conservative estimates. Moreover, India will remain totally dependent on the USA for spares, upgrades of engines and documents, aeroengine defect investigations, or major Repair and Overhaul/Maintenance, Repair and Overhaul (ROH/MRO) needs during the entire service life (30-35 years) of the aircraft and aeroengines. The conservative cost of MRO can be estimated at 50 per cent of the new procurement cost, i.e., US$ 3.09 billion. Thus, the life cycle cost of aeroengine procurement and MRO of GE aeroengines for indigenous fighter aircraft as per the 2021 cost reference will be US$ 9.26 billion, excluding mandatory or non-mandatory spares, consumables and futuristic upgrade costs.

In the commercial aircraft segment, India is one of the fastest-growing countries in the world in terms of passenger air travel, with more than 188 million passengers in the year 2022 (post-COVID-19), which include 22 million international passengers. India stands in the fourth position in terms of overall air passenger traffic (domestic and international); however, in terms of domestic air travel, India stands in the third position, behind the United States of America with 736 million, and China with 436 million passenger traffic per year, as per data published by the Sydney-based aviation think-tank CAPA-(Centre of Aviation) report. Presently, there are more than 400 airports in India; however, only 153 are operational ones. Many of these operational airports are linked to regional air connectivity through the UDAN project. India, at present, has approximately 716 commercial aircraft, out of which 475 are A319, 320, and A321; 84 are Boeing 737 aircraft; 51 are Boeing 747, 777, and 787 double aisle aircraft; and the remaining are turboprop ATR and Bombardier Q-400. As per data released by Airbus in respect of commercial aircraft demand in India on March 24, 2022, India requires 2,210 aircraft (1,770 small aircraft and 440 medium and large aircraft) in the next 20 years. As per general norms, commercial airlines may require 5-10 per cent reserve engines to cater for scheduled MRO of aeroengines; bird hits, foreign object damage, or other defects. Thus, India may require more than 4,420 new engines in the next 20 years for its commercial aircraft, and a similar number due to MRO. Presently, the approximate cost of the CFM LEAP engine installed on the A320 and B737 aircraft, is US$ 14.5 million and the MRO cost approximately 30-35 per cent of the new engine cost. Thus, India will spend approximately US$ 128 billion on the purchase of new commercial aircraft aeroengines in the next 20 years and approximately US$ 38.4 billion on the MRO of the engines in the next 20 years. As Indian civil aviation is witnessing unprecedented growth in terms of the number of domestic as well as international passengers and Air India has been taken over by Tata (an Indian private sector company), the company management announced at Aero India-2023, the purchase of 400 single-aisle aircraft (210 Airbus 320, 321 Neo aircraft, and 190 Boeing 737 Max family aircraft) and 70 twin-aisle commercial airliners. In order to power the single-aisle aircraft, Air India intends to buy 800 LEAP engines from CFM International (a 50:50 joint venture of GE, USA, and Safran Aircraft Engines, France) at an approximate cost of US$ 11,600 million, or 11.6 billion. Thus, in the civil segment, India will spend US$ 139.6 (128+11.6) billion in the next 20 years for the purchase of new aeroengines and MRO of aeroengines, whereas in the military segment, it will spend US$ 9.26 billion in the next ten years for the purchase of new aeroengines and MRO. The cost of MRO for all these aeroengines has been estimated at 35 per cent of the new aeroengine cost. Recently, another Indian airline operator, Indigo, announced placing orders for 500 aircraft of the Airbus A320 family on June 19, 2023, which will require 1,000 aeroengines of the CFM International LEAP class or equivalent, such as the P&W 1100G.

All the aircraft in the military and civil commercial segments so far have been procured from abroad, and the aeroengines installed on these aircraft were from the countries of the respective aircraft manufacturers. There is only a select group of countries with infrastructural facilities, design, development, and manufacturing capabilities for jet aeroengines with more than 2,000 kg of thrust, namely, the USA, Russia, France, and the UK. There is also an example of a successful collaborative aeroengine design and development known as the international aeroengine V-2500 for the single-aisle aircraft Airbus A320, McDonnell Douglas MD-90, and Embraer C-390 Millennium, designed, developed, and manufactured by a consortium of the following companies:
- Rolls-Royce – 12-stage HP compressor.
- P&W – combustion chamber and two-stage air-cooled HP turbine.
- Japanese Aeroengine corporation – fan and LP compressor.
- MTU Aeroengine – five-stage LP turbine.
- Fiat Avio – gearbox.

More than 7,600 aeroengines (V-2500) have been produced so far, and the engine has flown more than 200 million hours with more than 3,100 commercial aircraft in service around the world. This is a classic example wherein a high bypass turbofan aeroengine with different modules and major parts has been designed, developed, and manufactured by different countries, and the aeroengine has been assembled and tested by Rolls-Royce at a facility in Dahlewitz, Germany, and by P&W at its facility in Middletown, Connecticut. It is pertinent to note that the following companies have once again collaborated to design and develop the next-generation aeroengine, the PW 1100G-JM (geared turbofan), as a replacement for the V2500 international aeroengine.

*   P&W, USA.
*   MTU, Germany.
*   IHI, Japan (a member company of the Japanese Aeroengine Corporation).

The success of the V-2500 and next-generation PW 1000G-JM has been attributed to a collaborative arrangement, hard work, expertise, experience, mutual trust, respect for each other, and ample communication. Based on the lessons learnt from this collaborative approach, it is possible to adopt this model for indigenous development in India for low bypass turbofan core engines of 110-130 KN thrust under a collaborative arrangement on a risk and revenue sharing basis between private and public sector enterprises with initial developmental project funding from the Government of India.

The following companies are prominent aeroengine and aeroengine major component designers and manufacturers:

*   Rolls-Royce, UK.
*   GE, USA.
*   P&W, USA.
*   Safran (erstwhile Snecma), France.
*   CFM International – a joint venture of GE and Safran, France.
*   Engine Alliance – a joint venture of GE and P&W.
*   Honeywell, USA.
*   United Engine Corporation, Russia.
*   Motor Sich, Ukraine.

Major aeroengine parts designers and manufacturers are:

*   MTU, Germany.

- Japanese Aeroengine Corporation, Japan.
- Fiat Avia, Italy.
- Ishikawajima-Harima Heavy Industries (IHI), Japan.
- GE Honda Aeroengine, Japan.

In India, HAL is the only agency engaged in licensed manufacturing of aeroengine components, assembly, and testing, and MRO of aeroengines for United Engine Corporation, Russia; NPO Saturn, a subsidiary of United Engine Corporation, Russia; and Rolls-Royce, UK.

As evident from the list of global aeroengine manufacturers, there are primarily only five OEMs, designers, and manufacturers of high-thrust turbofan aeroengines in the world, and the rest are their own joint ventures raised primarily for a specific engine project's design, development, and manufacturing or for the manufacturing of major modules. The five primary aeroengine designers, developers, and prototype manufacturers are as given below:

- Rolls-Royce, UK.
- GE, USA.
- P&W, USA.
- United Engine Corporation, Russia.
- Safran (erstwhile Snecma), France.

GTRE, Bengaluru, is keen on forming a joint venture or an agreement for technological collaboration for joint co-development of an aeroengine with one of the established aeroengine designers or manufacturers for a low bypass turbofan aeroengine of 110-130 KN thrust that can be installed on India's ambitious fifth generation AMCA. It is understood that GTRE is in the advanced stages of negotiation with Safran, France; Rolls-Royce, UK; and GE, USA. However, the domain of aeroengine design, development, and manufacturing of turbofan aeroengines is a very niche segment of technology. All the aeroengine manufacturing companies and countries have spent a prohibitive amount of resources in mastering the technology and, thus, do not want to proliferate the technology in view of their huge economic receivables from the sale of new engines and the upkeep of the aeroengines already under operation with various operators globally, which they have supplied to other countries in terms of providing spares, test equipment, upgrade of technology, life extension, and MRO services. Hence, they are completely against generating their own competitors for future market opportunities. Therefore, in all probability, none of the five major aeroengine designers or manufacturers is likely to agree to any joint venture or co-developing an aeroengine at any cost or consideration. At best, they may agree on a specific aeroengine ToT, licensed manufacturing, or a conditional agreement for joint development for the design and development of a few specific modules, systems, or sub-systems at the lower end of the technology segment and continuously supply the critical aeroengine modules at the higher end of the technology from their own company or country. These aspects need thorough debate, understanding, and tough negotiations from the Indian side. There is likely to be a host of other conditionalities, such as:

- No futuristic improvement of aeroengines without the OEM's permission.
- No sale of aeroengines to any third country.
- No life extension programme without their permission.
- No use of technologies co-developed for any other military or civil aeroengine segment.
- No alternative, replacement, or indigenous substitution of any critical module, system, or aggregate which are being supplied by foreign companies.
- Non-transparency or non-involvement of Indian specialists for defect investigation and analysis in case of a major failure, incident, or accident.
- Non-use of Indian raw material, even if it conforms to aerospace-grade material.
- Non-use of Indian fasteners, cable connectors, cable looms, and consumables, making India dependent on the source of supply for the entire service life span of aircraft or aeroengines.

- Non-usage of Indian tools, trollies, zigs, fixtures, and ground equipment.
- Non-usage of the cut engine for training or educational purposes.
- Insistence on sharing operational exploitation data of aeroengines with the pretext of reliability studies and improvement programs for their own usage.
- MRO of the engine at their designated agencies.
- Insistence on payment of royalty on each aeroengine in production, future upgrades, and updated documents.
- Insistence on the supply of consumables such as engine oil from the OEM's sources of supply.
- Non-sharing of data pertaining to flight and maintenance safety experienced by their own Air Forces.

GTRE, Bengaluru, is a research institution under the DRDO, Ministry of Defence, and has the same level of functioning, bureaucratic approach, and restrictive business practices as other public sector undertakings of the Government of India. When we compare the general functioning of a typical public sector company with that of a private sector enterprise on some of the following common parameters, the public sector units appear to be less efficient than private sector enterprises in general:

Sl. No.	Parameters	Public Sector Company	Private Sector Company
1.	Decision-making process	Slower	Faster
2.	Time schedule for task execution	longer	Shorter
3.	Project execution cost	Higher	Lower
4.	Bureaucratic hurdles	Higher	Lower
5.	Product quality	Good	Very good
6.	Technology absorption rate	Slower	Faster
7.	Dynamism	Normal	Good
8.	Workers' compensation	Good	Normal
9.	Rate of return on capital	Normal	Very good
10.	Procurement process	Sluggish	Very fast
11.	Inventory management system	Normal	Very good

12.	Upkeep of machines and work environment	Normal	Good
13.	Adherence to laid down rules	Good	Good
14.	Adherence to environmental protection	Good	Normal
15.	Maintenance of record (Traceability)	Good	Very good
16.	Degree of commitment	Moderate	Excellent
17.	Pay, perks and remunerations	Same pay and perks for the same cadre of staff, whether efficient or inefficient	Different pay and perks for the same cadre of staff, based on proficiency

Although the parameters and the assessment of various attributes in the table above are arbitrary and presumptive and may not be applicable to all public or private sector enterprises, they may find applicability in a large segment, including Defence Public Sector Undertakings (DPSUs) and large successful private sector enterprises in India.

It is important to note that India has made tremendous progress in the space sector, including the Chandrayaan-3 launch recently by the Indian Space Research Organisation (ISRO). ISRO's success can be primarily attributed to a very high degree of commitment from the noted scientists, and absence of bureaucratic hurdles. ISRO has successfully developed a large range of materials (metals as well as composites) indigenously with active support from academic institutes, Public Sector Undertakings (PSUs), and private sector companies. GTRE can probably follow a similar model and involve nationally reputed academic institutes and the concerned research institutes, including the Council of Scientific and Industrial Research (CSIR) laboratories. GTRE did not succeed in its objective to develop the Kaveri engine due to varied reasons that may or may not be attributable to GTRE directly.

It is pertinent to note that the requirement of a flight test-bed was not envisaged by GTRE at the beginning of the project, and no attempt was made to develop a flying test-bed either on a temporary or permanent basis; hence, our dependence on a flight test-bed abroad was by default. It is important to note that a flight test-bed is generally a multi-engine transport aircraft, specifically modified for the intended function of research and testing for the installation of a new aeroengine on a specific engine mounting point and hard-wired to obtain all the engine running parameters for analysis and evaluation of the engine. India should earmark a partially used Boeing 747 or Ilyushin Il-76 aircraft and attempt to make a flight test-bed indigenously. P&W has two Boeing 747 SP aircraft test-beds for testing the experimental and conceptual aeroengine designs before entering manufacturing. Thus, it can be clearly seen that all the major aeroengine manufacturers have their own flying test-beds for carrying out full-fledged altitude testing, aeroengine research, and performance improvement programmes. All the parameters that are monitored and recorded at the ground engine test-bed need to be monitored at an altitude of 30,000-40,000 ft for validation of the engine's performance. India has adequate experience in the design, installation, and integration of ground engine test-beds with various sensors, data transfer integration, and recording and display systems. Hence, it is feasible to undertake a project for the indigenous design and development of a flight test-bed. It is a worthwhile proposition to consider the design and development of a flying test-bed for futuristic aeroengine testing in India. It is prudent to note that most of the parameters can be evaluated based on the ground testing of the aeroengine test-bed; however, in order to evaluate the engine for the altitude test and validate the ground test results, it is necessary to have a flight test-bed. It is important that India go for an independent programme for a flight test-bed, preferably on a Boeing 747 or Ilyushin Il-76 aircraft, indigenously, irrespective of the indigenous design and development of the 110-130 KN core aeroengine. The indigenous flying test-bed will be a big boon for expediting the pace of development of indigenous aeroengines and also for the performance testing and improvement of existing aeroengines. It is preferred to entrust the task of design and development to a reputed private sector engineering company and the certification task to the Centre for Military Airworthiness and Certification (CEMILAC), Bengaluru, to ensure time-bound completion of the project.

A comprehensive study on aerospace-grade metals needs to be undertaken on priority, preferably by an industry association (CII, SIDM,

FICCI, PHD Chamber of Commerce and Industry, etc.) to assess the requirement of various equivalent-grade materials for aeroengine development and the capabilities of the Indian industries to produce the material indigenously. A problem may arise due to the insufficient quantity of material required for the aeroengine's design and development. It may not make economic sense to develop the material for a private company. Hence, it is strongly recommended that the government make the necessary expenditures for material development. Mishra Dhatu Nigam Limited (MIDHANI), Hyderabad, a Government of India enterprise, is fully capable of developing a large range of steel, titanium, and superalloys with research assistance from the Defence Material Research Laboratory, Hyderabad. It is important to note that for light materials, the Jawaharlal Nehru Aluminium Research Development and Design Centre (JNARDDC), Nagpur, may be consulted for the development of aerospace grade aluminium such as Russian grade D-16/D-17/D18/D-19 or US grade equivalent aluminium Al-2024/Al 7085 in India. The processing of aerospace-grade aluminium should preferably provide the following minimum characteristics:

* Tensile strength   =   290 MPa
* Yield strength   =   172 MPa
* Relative elongation   =   25 per cent
* Brinell hardness   =   823PMa

It is important that the expenditure on the development of various types of aerospace-grade materials be undertaken by the Government of India as a strategic investment in order to reduce or eliminate our continued dependence on foreign countries. The development of indigenous aerospace-grade materials will go a long way in the design and development of indigenous aircraft and aeroengine, reduce delays caused by the import of materials and the high cost of imports, and eliminate our dependency on foreign countries.

Adequate efforts should be made for the development of composites in India since the latest trend in the aerospace segment is also moving towards enhanced use of composites to reduce aeroengine weight, have a longer operational life, maintain good strength, and reduce corrosion effects. Although a number of companies have entered the composite manufacturing segment in India, it is essential to establish a national research laboratory on composites to lay down the standards, undertake niche research on various types of fibres, polyesters, high-strength resins, adhesives, and prepregs, and advance digital large-size autoclaves for thermal processing, curing, and bonding of various parts for futuristic usage on aircraft and aeroengine applications.

Design and development of tools, trolleys, dies, zigs, and fixtures are considered extremely important for indigenous design and development of aeroengines. It is important to note that the Government of India, through the Ministry of Micro, Small and Medium Enterprises, has established a number of tool rooms and technology centres in various parts of the country for different segments such as general engineering, electrical and electronics component design and manufacturing, calibration and testing, hand tools, foundries and forge, the glass industry, sports goods, fragrance and flavour, and footwear. These tool rooms and technology centres are well-equipped and capable of developing modern tools, dies, and fixtures. The relevant tool rooms, which can be of immense help to the aeroengine development agency, are as given below:

Sl. No.	Place	Name of Technology Centre	Sector Serving
1.	Bhubaneswar	Central Tool Room and Training Centre (CTTC)	General Engineering (aerospace precision components)
2.	Jamshedpur	Indo-Danish Tool Room (IDTR)	General Engineering (auto parts-metal)
3.	Kolkata	Central Tool Room and Training Centre (CTTC)	General Engineering
4.	Guwahati	Tool Room and Training Centre (TRTC)	General Engineering
5.	Aurangabad	Indo-German Tool Room (IGTR)	General Engineering (auto parts-metal)
6.	Mumbai	Electrical and Electronic Components Design, Manufacturing, Calibration and Testing	Electrical and Electronics components, looms, wirings, Connectors, test equipment calibration
7.	Indore	Indo-German Tool Room (IGTR)	General Engineering (auto parts and pharmaceutical-plastic)

8.	Ahmedabad	Indo-German Tool Room (IGTR)	General Engineering (auto parts and plastic components)
9.	Ludhiana	Central Tool Room (CTR)	General Engineering (auto parts–metal)
10.	Hyderabad	Central Institute of Tool Design (CITD)	General Engineering (automation)

All the tool rooms and technology centres have state-of-the-art facilities, including precision engineering machinery, to develop any type of tool, die, zigs, or fixture with engineering design digital records. For subsequent orders, only the design number is sufficient for future manufacturing.

Glue and Adhesive for High-Temperature Operation

One of the most critical requirements for an aeroengine is to have the abradable coating in the respective engine casing slots where the compressor and turbine rotating blades' tips are rotating. The development of abradable coatings and other glues and adhesives is considered extremely important since these items have a very short shelf life, and sometimes the life of the items, if imported, expires by the time it reaches the intended user. It may be noted that the gap between the rotating blade tip and the engine casing should be ideally zero, wherein the rotating blade tip should make a groove in the abradable material. The abradable material should have excellent shear strength. It should not crack or chip off in the compressor segment, and the respective material used in the turbine segment should be able to withstand high temperatures. The material should be softer than the blade material to avoid damage to the blade tip in the case of any minute thermal elongation due to prolonging the operation of the aeroengine at high temperatures. In India, the Council of Industrial Research (CSIR)-Indian Institute of Chemical Engineering, Hyderabad, has been closely associated with ISRO for a number of products' development for space applications. The institute has the capability to develop high-strength bonding ceramic glue with abradable characteristics for operation at high-temperature ranges. The Indian Institute of Chemical

Technology (IICT), Hyderabad, undertakes a large number of research projects in the basic and applied domains of chemistry, biochemistry, and chemical engineering and has filed the maximum number of Intellectual Property Rights (IPRs) among the CSIR laboratories. The institute can contribute immensely to the development of various chemical products for indigenous aeroengine development. The air gap between the rotating turbine blades and the static engine casing should be maintained as low as possible so as to reduce air leakage, which results in engine inefficiency. This action is similar to that of the compressor. The abradable material is also coated inside the static turbine casing or the metal ring in some of the aeroengines, while a number of designers use a number of shroud segments made of softer material in the honeycomb format as a sealing mechanism. Unlike the compressor segment coatings, the turbine segment static part coating must withstand high temperatures to the tune of 1000°C and above. A special coating material is used, which is an aluminium-silica-based alloy named yttria, ytterbium oxide (ytterbia), or Yttria Stabilised Zirconia (YSZ), porous ceramics, or an alloy of softer ceramic metals with a base coat and a harder material as a top coat with different characteristics. YSZ is the most popular coating, generally coated with plasma spray machines for turbine blades. It provides wear and corrosion resistance in addition to withstanding thermal stresses. There are some other specialised metal powders available for use as bond coats and top coats for use as Thermal Barrier Coatings (TBCs); however, the suitability of base adherence and wear characteristics needs to be studied carefully. It is important to note that a number of coating and wear-resistant corrosion protection slurry options are available for aeroengine parts, such as the Sermeloy-J, Serme-T, and Sermetal-W, and already in use; however, the selection is based on the substrate metal characteristics and chemical compositions. The coating material should have the specific characteristics of good galvanic corrosion resistance and oxidation resistance of at least 1000°C. The turbine disc and the blades are subjected to cyclic heating and cooling, thereby causing high operational fatigue in addition to various forces acting on them and causing additional stresses. The size of the turbine blades from the first stage to subsequent stages keeps increasing since the hot gas' pressure reduces in each stage. Hence, a higher turbine blade contact surface area is required to extract power from the relatively lower pressure gases and to handle a large volume of gases. The military aircraft aeroengines are also incorporated with an afterburner mechanism to burn additional fuel in the jet pipe to raise the gas temperature and obtain additional jet velocity. It is possible to achieve up to 50 per cent additional thrust and, hence, help in the reduction of aeroengine size and weight for military aircraft.

Materials and Manufacturing of Inserts, Spacers, Struts, Fasteners, etc.

It is prudent to note that special materials and processes are used for the manufacturing of inserts, spacers, struts, and various shapes and sizes of fasteners for the integration of various modules or sub-sections of the aeroengine. These are designed to withstand high temperatures and high stresses during the operation of the aeroengines. A number of Indian companies, such as TVS, Ankit Fasteners, and Deepak Fasteners are manufacturing world-class mating tools such as inserts, spacers, struts, and fasteners and supplying them to foreign-based OEMs for aviation usage.

Main and Auxiliary Gearbox

All jet engines must necessarily have a gearbox to transfer power efficiently from the aeroengine to various mechanical, electrical, pneumatic, and hydraulic systems of the aircraft and even to some of the systems of aeroengine controls. The smaller aeroengines have only a single gearbox. In contrast, the bigger aeroengines generally have two gearboxes mechanisms to convert the mechanical rotational energy of the turbine shaft to electrical, pneumatic, and hydraulic power sources through electrical motors (DC and AC), compressors, and hydraulic pumps (main and standby). The provision of a dual gearbox also provides inherent system redundancy and much higher operational reliability. The various types of gears, such as spur, rack and pinion, helical, rotor, worm and worm wheel, mitre, screw, herringbone, and bevel, are prominently used in the aeroengine gearbox depending on the type and role of the aeroengine, e.g., in the turboshaft engine used for the helicopters, the gearbox's main role is to transfer the entire power of the engine to the main rotor and tail rotor (or a tandem rotor helicopter, which has two main rotors and no tail rotor) through reduction gears placed in a sun and planetary gear arrangement in one or two stages of speed reduction and fractional power to the electrical motors for generating electrical power and to the hydraulic pump for providing a hydraulic power source. Similarly, in turboprop jet engines, almost the entire power of the turbine rotor is transferred to the propeller through the front gearbox, which has a sun and planetary gear arrangement for single or dual stages of speed reduction. Many companies are engaged in the manufacturing of various types of gear in India for several industrial applications. These companies are capable of designing and prototyping aviation grade gears and can be assigned the manufacturing of aeroengine gears in India. Some of the prominent gear manufacturing companies in India are as follows:

- Gear and Gear Drives (India) Pvt Ltd, Bengaluru.
- Anand Udyog, Belgaum.
- SAN Engineering and Locomotive Company Limited, Bengaluru
- Rotec Transmission (P) Ltd, Medak.
- Shanthi Gear Ltd, Coimbatore.
- Automotive Axles Limited, Mysore and Jamshedpur.
- Essen Aluminium, Belgaum.
- Hind Gears, Kolhapur.
- Mess Well, Bengaluru.
- Orients Plants, Coimbatore.

All these companies listed above are already manufacturing different types of precision gears for a plethora of applications in various industrial segments in India and abroad. However, for the aviation application, the specific requirement is the highest reliability, and minimum wear and tear in order to ensure the safety of the operation. It is essential to incorporate special processes on the gears post-manufacturing, depending on the type of material used, such as case hardening, in order to increase the hardness of the outer surface while keeping the core or centre unaffected by the processing of the metal. As prominently used in the industrial segment, the following processes are used for the case hardening of a metallic gear:

- **Carburising:** It is one of the most popular processes used for case hardening, wherein the material to be case hardened is exposed to a carbon-rich environment such as charcoal and heated at a temperature range of 790°C to 925°C for a specified time-frame to achieve the desired hardness.

- **Nitriding:** It is also used for strain hardening in a similar way as carburising, with the material exposed to nitrogenous environments such as ammonia. Nitriding is more popularly used for aviation-related parts since the environmental temperature and amount of nitrogen to be flown in the furnace can be more precisely controlled through a computerised furnace control system.

Compressor and Turbine Rotor

As covered earlier, India has become an important hub, a centre for the supply of automobile spare parts to almost all the major automobile companies in the world. Some of the global automobile manufacturers have also shifted their manufacturing base to India, thereby creating world-class engineering product manufacturing infrastructure in the private sector. Many companies have attained adequate skill and expertise to design, develop, and manufacture aviation-related precision engineering mechanical and electrical parts in India. Some of the Indian industrial enterprises are already engaged in the machining and manufacturing of aviation critical spare parts and supplies to global aircraft and aeroengine manufacturers across the world. However, most of these companies are presently working only as manufacturers as per the build to print philosophy and not actually designing the spare parts. It is pertinent to note that India has a vast pool of engineers and academicians who can be engaged in the design process of the critical and complex parts of the aeroengine. Presently, with the advent of a large number of design and simulation tools available, it is possible to validate the design process well in advance before the parts are actually produced, thereby reducing the cost and the time-frame of the design and developmental process. The real challenge is to lay down the specifications of the product for its intended function, for which the users have to work meticulously, keeping some scope for the futuristic improvement's provisions. Some of the Indian automobile companies are already manufacturing power transmission shafts for Mercedes-Benz and other reputed car manufacturers abroad as per the designs provided by the car manufacturers. The process expertise and the necessary wherewithal are available; however, we must orient these manufacturers to upgrade their facilities for the machining of compressor and turbine rotors through numerically controlled servo mechanism systems for even better precision.

The compressor or turbine rotor is machined out of forged steel grades 25 Cr1Mo1VA, 26CrNI3Mo2VA, or 23CrMoNiWV88, and steel X750. The turbine rotors are generally machined through turning and groove-cutting machine operations. The turbine rotors are also subjected to drilling to make internal core passages for oil lubrication of the support bearings. The turbine blades and the turbine discs are generally made from nickel alloys since they provide adequate strength, hardness, corrosion protection, and wear resistance. It is important to note that the Japanese scientist Professor Kyosuke Yoshimi of Tohoku University's Graduate School of Engineering has recently developed a new metal that can prove to be better than nickel-dominated superalloys for aircraft and aeroengine applications. It is understood that the new alloy is a titanium carbide (TiC)-reinforced, molybdenum-silicon-boron (Mo-Si-B)-based alloy, or MoSiBTiC, whose high-temperature strength was identified under constant forces in the temperature ranges of 1400-1600°C which is most suitable for futuristic jet engine turbines, and even for electric generation steam turbines. Presently, the nickel-dominated superalloys suitable for manufacturing turbine shafts are not manufactured in India, so they have to be imported. However, all efforts should be made to develop this superalloy in India in order to reduce the cost and lead time of imports and faster development of aeroengines in India.

Afterburner

The military fighter and combat aircraft aeroengines necessarily use an afterburner system to obtain higher thrust during take-off and combat operations in order to avoid bigger and heavier aeroengine installations on the combat aircraft and also to have a higher degree of manoeuvrability. The afterburner arrangement is integrated into the aeroengine after the LP turbines in the jet pipe. It is generally in the form of three concentric hollow circular rings supported rigidly with fairings and studs from the jet pipe structure. The afterburner rings get high-pressure fuel from the afterburner fuel pump connected from the base to the rings. The afterburner rings have uniform fine jets to spray fuel in the jet pipe, which gets partially atomised due to collusion with high-speed exhaust gases from the turbine. The exhaust gas from the turbine and the bypass air flowing over the core engine from the fan duct mix with the afterburner fuel in the jet pipe. It burns with an elongated flame, and, in turn, it heats up the gases to a very high temperature (up to 1700°C), and, thus, the higher expansion of the hot gases results in very high jet velocity. An efficient afterburner system such as the F-135 aeroengine installed on the F-35 combat aircraft variants can provide up to 50 per cent additional thrust to the aeroengine, which is generally known as 'wet thrust.' A provision is also made to control the jet nozzle's partial or full opening or closing, depending on the mode of operation of the aeroengine in terms of the pilot's throttle setting position. However, in the case of active afterburner operation, the jet nozzle remains fully open to cater for the maximum air velocity or generation of the highest thrust.

Full Authority Digital (Electronic) Engine Control (FADEC)

FADEC was experimented with by a combined team of NASA and P&W on the TF-30 aeroengine installed on F-111 aircraft in 1970. The full authority analogue electronic control was also experimented with on the Rolls-Royce/Snecma Olympus 593 engine installed on the supersonic transport aircraft in 1960. The development started with hydro-mechanical control and electronic-supervisory control and was subsequently replaced by analogue electronics control and then a digital electronics control system. FADEC receives input from all sensors employed for monitoring the aircraft's flying parameters, including the throttle position, and optimises the fuel input to the aeroengine through the control unit of the main fuel pump. FADEC also controls the jet nozzle opening and closing of the aeroengine, aircraft stall limits, vertical and horizontal gravitational force limitations, rate of climb, rate of descent, etc., under various modes of aircraft operation. It is feasible to develop a FADEC indigenously based on the design of engine operating algorithms for the engine operation at various settings by experienced maintenance engineers with adequate domain knowledge and the design of suitable software code for the fuel inlet control by software experts. A typical FADEC aeroengine control system's design and development include the following aspects:

- Electronic circuit design.
- Ruggedised FADEC enclosure design (FADEC box).
- Controller EMI/environmental testing.
- Software/hardware integration and test bench testing.
- System integration and engine test support.
- Generating control system specifications.
- Control algorithm development.
- Real-time FADEC software development.

The design and development of FADEC must include some of the essential features, such as built-in-test, diagnostics, and prognostic aspects. The data from and into the FADEC should be transmissible through various communication bus bars, such as CAN (ISO 11898), ethernet, USB, and MIL-1553. The following qualification tests are essential for FADEC certifications by the relevant authorities:

- Corrosive environment (salt fog).
- Salt water immersion.

- High/low temperature.
- Temperature shock.
- Humidity.
- Vibration.
- Altitude.
- General shock.
- Explosive environment.
- EMI suite.
- Electro-static discharge.

These tests must conform to FAA or Military Standards MIL-STD-810 and MIL-STD-461.

10

Proposed Way Forward

The great initiative of Prime Minister Narendra Modi known as *Aatmanirbhar Bharat Abhiyaan*, or self-reliant India, is considered extremely important and promises to deliver far-reaching results in the future, especially pertaining to the requirement for war-waging machines for the Indian armed forces. It is extremely important to note that India cannot fight its future wars with foreign purchased war-waging machines if it is not aligned with those very countries that have supplied the military hardware. However, suppose India decides to overrule, and still dares to go to war without alignment with, the countries that have supplied it with critical military hardware. In that case, the consequences can be disastrous since India can be struck off from the global critical spare parts supply chain. The sustenance of our war-waging capabilities will be extremely difficult.

Air power is one of the most important tools available to the government to project power in the international arena and achieve unexpected, out-of-proportion results most swiftly. The examples of the Balakot air strike and the airlifting of Indians from war-struck countries such as Kuwait, Sudan, and Ukraine in the recent past are considered appropriate to understand the importance of air power. Air power, besides being flexible, also provides the much-needed surprise element of global reach. Air power will remain the most dominant factor in winning future wars with its four fundamental elements such as intelligence and situational awareness, air mobility, offensive action, and achieving air superiority.

India has taken a great step towards indigenous design, development, and manufacturing of the LCA Tejas with an aeroengine from GE, USA. It is important to note that the aeroengine is one of the most crucial machines powering the aircraft, besides providing electrical, mechanical, and hydraulic power to the aircraft.

India's experiment with an indigenous aeroengine was marked with the Kaveri aeroengine development programme, sanctioned by the Government of India in March 1989. However, the project failed to provide any substantial contribution to India's aspirations.

In contrast to the accomplishment of GTRE under DRDO, ISRO has made tremendous progress in the space segment, including the design and development of satellites and launch vehicles, cryogenic engines, and a plethora of related indigenous technologies to the field of space exploration and space-based applications. A large variety of sensors, payloads, and space-grade materials have also been developed by ISRO indigenously through a large network of Micro, Small and Medium Enterprises (MSMEs), research establishments, and academic institutions of repute. The design and development of the Polar Satellite Launch Vehicle (PSLV) and Geostationary Launch Vehicle (GSLV) resulted in putting satellites into polar and geostationary orbits. These rockets helped ISRO launch a large number of communication satellites, earth observation satellites, missions to the moon (Chandrayaan-1, Chandrayaan-2, and Chandrayaan-3), and the Mars Orbital Mission.

There are many reasons for ISRO's success in the space sector; however, some of the most prominent can be attributed to the following:

- Strong leadership and a high degree of professional commitment.
- Independent organisations functioning directly under the prime minister.
- No bureaucratic hurdle or intervention.
- Cost-effective technology developments.
- Timely availability of developmental fund.

The ISRO chairman is also the chairman of the Space Commission and the secretary of the Department of Space. This arrangement has ensured vertical integration with the policy-makers, who clearly understand the nature of the various projects to be undertaken by ISRO and the likely benefits the country is likely to derive from the specific projects.

The aeroengine will remain the most relevant machine as the power plant for the military as well as civil, commercial and private aircraft for at least the next 40 to 50 years until it is possibly replaced with ionic wind thrusters or electrical ion thrusters, for which research is in progress at some of the top technical institutes in the world. Ironically, even after 75 years of its independence, India is 100 per cent dependent on the advanced countries (the USA, UK, France, and Russia) for the supply of aeroengines, spare parts, and consumables for the sustenance of its military aircraft and even the commercial and private jet aircraft of civil aviation. Total reliance on these countries for India's aviation sector sustenance in the future is considered highly vulnerable. It may paralyse the country in the case of any future sanctions imposed by any of these countries in the fast-moving polarisation and geopolitical environment.

In view of this backdrop, it is strongly opined that a low bypass turbofan aeroengine core of 110-130 KN of thrust development must be initiated and completed within a time-frame of five years on top priority in an accelerated mission mode. The same core of the aeroengine is to be utilised for the military combat fighter aircraft, as well as single aisle civil and commercial aircraft, and the following five options are considered relevant for the time-bound progression of the Indian indigenous aeroengine developmental programme:

1. Aeroengine development by GTRE with Transfer of Technology (ToT), joint venture, or co-development with established OEMs.
2. Fifty-one percent divestment of GTRE and aeroengine development by GTRE.
3. Hundred percent sell off of GTRE to an Indian private company, and aeroengine development by a new entity.
4. Aeroengine development by public-private partnership on a collaborative approach through risk and revenue sharing basis.
5. Aeroengine development in India in mission mode under the Prime Minister's Office (PMO) in line with ISRO.

Aeroengine development and its related technologies are capital intensive and it may be difficult to find adequate private sector companies to undertake investment in its research, design, and development. The merits and demerits of all the above-mentioned five options can be studied and deliberated upon, and a final decision may be arrived at for fast-track development of dual-utilisation (civil and military) low bypass turbofan aeroengines. The Ministry of Defence and Ministry of Civil Aviation should preferably participate in the programme jointly since both ministries will derive final benefits pursuant to the successful development of indigenous aeroengines in India. Now we shall discuss and debate all the merits and demerits of each of the five options as follows:

Option-1

Aeroengine Development by GTRE with ToT, Joint Venture, or Co-development with Established OEMs

Aeroengine development by GTRE through ToT from OEMs, a joint venture with credible OEMs, co-development or joint development is indeed possible but access to critical technologies must be provided to GTRE. Aeroengine design and development activities are challenging, however, these can be undertaken through ToT, co-development, or joint development with the help of a foreign-based established OEM.

It is pertinent to note that the established aeroengine OEMs abroad have incurred huge expenditures in terms of revenue and time to develop and mature the technology; hence, under no circumstances will they part with their complete technology (100 per cent of technology), including the design data. It is understood that GTRE, Bengaluru, is already engaged in discussions with Rolls-Royce, UK; Safran, France; and GE, USA.

The merits and demerits of this option are enumerated below:

- The critical aggregates or modules of the engine technology may not be provided since the OEM of some specific aggregates or modules may be different from the OEM of the aeroengine integrator and may also be from another country.

- Some of the modules or units will be supplied by the OEM, resulting in partial dependence on the OEM for that module or unit for the entire life cycle of the aeroengine.
- Restriction on futuristic improvement of the aeroengine without the OEM's permission.
- Denial of sale of aeroengines to any foreign-friendly country.
- No life revision or extension programme is allowed without the OEM's permission.
- Non-use of technologies co-developed for any other military or civil segment.
- No replacement or indigenisation of any critical module being supplied by foreign companies.
- Non- transparency or non-involvement of Indian specialists for defect investigation and analysis in case of a major failure or accident in the future.
- Non-use of Indian raw materials even if they conform to aerospace-grade material standards.
- Non-use of Indian fasteners, cable connectors, cable looms, and consumables make India dependent on their source of supply for the entire life span of an aircraft or the aeroengine.
- Non-usage of Indian tools, trollies, zigs, fixtures, and ground equipment.
- Non-usage of the cut engine models for training or educational purpose.
- Insistence on sharing operational data.
- MRO of the engine at designated agencies only.

As already covered, GTRE has spent adequate time (33 years) and resources (more than ₹2,100 crore approximately) on the Kaveri aeroengine design and development. The engine could not generate adequate thrust and could not meet the total weight criteria for the LCA Tejas, and the country could not derive any benefit from this project, directly or indirectly. It may not be prudent to continue a fresh developmental phase of aeroengine design and development with GTRE without incorporating any cultural and structural changes in terms of more autonomy, a strict time schedule, a hire and fire approach, attracting talents at higher remuneration, and proper accountability. The negotiation with the established OEM should not impose any unreasonable restrictions that can affect our right to apply the knowledge to other Indian projects (civil or military) in the future or restrict us from selling the finished product to any foreign-friendly country in the future. Presently, under the prevailing political environment in India, it may be extremely difficult for the government to incorporate a specific work culture, adopt a hire-and-fire approach, attract talent at higher remuneration, maintain a strict time schedule, and fix accountability only for GTRE. Since the government is expected to be an ideal employer and cannot give higher pay to a certain superior class of talent, higher incentives, or even basic pay and perks, it may not take any hard decisions pertaining to a change of work culture or ease of the rigid hierarchy. Hence, this option does not seem to be the most suitable option for the design and development of a state-of-the-art indigenous aeroengine.

Option-2

Fifty-one percent Divestment of GTRE and Aeroengine Development by GTRE

In case the Government of India resorts to selling off its majority stake in GTRE to top Indian private industrial enterprises and the control of GTRE comes under proficient professionals or a group of professionals, it will be able to turn around the work culture, ethics, and strict schedule of the project time-line and even implement a hire-and-fire approach. This provision can make GTRE more efficient and productive and may be even eligible for foreign direct investment. Although the valuation of GTRE may not be attractive for the Government of India to opt for divestment of GTRE, the government can take an early decision pertaining to the labour law wherein there should not be any pre-condition pertaining to retention, promotion, or other terms and conditions of service during the divestment of the 51 per cent stake in GTRE and thereafter to Indian industrial enterprises. The advantage to Indian private industries will be the instant availability of an extensive knowledge base and technical knowhow at GTRE, along with a large number of related machinery and other infrastructures that are considered vital for commencing the design, development, prototype manufacturing, and ground performance test of the aeroengine test-bed at the functional engine test-bed at GTRE. Strict terms and conditions must be set, and the government must provide funds in terms of production linked incentives, which may be released strictly as per the progress of the development of the aeroengine in a pre-decided phased manner. The governing board should comprise experts with domain knowledge of aeroengines, and the Chief Executive Officer (CEO) should be dynamic and result oriented with a proven track record of managing engineering research organisations or an experienced person with strong hands-on experience. It is pertinent to note that perfect aeroengine design and development have not taken place in any country on the very first attempt. A number of iterations are done in the design and manufacturing phases and even after the ground and air test-bed runs based on the realistic parameters obtained vis-à-vis the original design documents. When the aeroengine is dismantled post-ground/ air test-bed run, it gives tremendous data for analysis and futuristic improvement by tweaking the design or the critical processes. Hence, the closure of the Kaveri aeroengine project without real learning value as to what went wrong or what the shortcomings were in terms of technology inadequacy, material availability, process engineering, or design shortcomings needs careful examination and wider discussion to ensure proper resolution. If we study and analyse the development of any civil or military aeroengine in the world, we find that incremental improvements continue during the entire life cycle of the aeroengine and a number of variants one created, and even the thrust is increased by the introduction of better-improved material, improved processes, enhancement of efficiency, or fuel consumption reductions.

If this option is chosen for the design, development, and prototype manufacturing of the low bypass turbofan engine, based on the experience gained by GTRE, it is essential to find out and understand the exact reason for the Kaveri aeroengine's failure and the likely ways and means to resolve the problem areas with the active support of practising professionals and academia from reputed institutes in India or abroad. However, there is every likelihood of Indian industrial enterprises shying away from participating in the divestment process of GTRE for the aeroengine development programme due to the heavy capital investment requirement, the existing work culture, the ethos of government organisations, and the long gestation period of the developmental cycle of the project owing to its complexities and the development of some of the critical technologies in the specified time-frame.

The changes in labour laws, the complexities involved in the government's funding of a private company, and the eagerness of private companies as major stakeholders may or may not succeed in the design, development, and prototype manufacturing of jet engines in India in the strict time-frame of five years or so. Thus, this option also does not appear to be the best one. However, in case this option is adopted for pursuing the aeroengine development, in order to make it a workable one, the government may like to create a favourable environment in terms of assured business potential opportunity besides special incentives and subsidies for critical technology development.

Option-3

Hundred percent Sell-off of GTRE to an Indian Private Company and Aeroengine Development by a New Entity

GTRE has the aeroengine domain knowledge, experience, and skill set for the design, development, and prototype manufacturing of low bypass turbofan core aeroengines, besides large machinery, state-of-the-art test equipment, an engine ground test-bed, and other related infrastructure. A 100 per cent sell-off of GTRE to an Indian private enterprise solely for aeroengine design and development of indigenous aeroengines may be one of the most viable options for expediting the indigenous aeroengine developmental process. The Government of India may consider selling GTRE completely to a private Indian entity that can become a facilitator for accelerated indigenous aeroengine development. The aeroengine technology is quite complex and challenging, and it requires huge capital investment for design and development. However, the private sector may or may not make such heavy investment without active government support for the funds as well as hand-holding for international certification and acceptance. While the Indian aeroengine may become successful in meeting the technical parameters and the safety and reliability aspects, the OEMs of the aircraft may still oppose the installation of the Indian aeroengine on their aircraft due to commercial reasons; hence, it may require the Indian government's intervention. The common practice in most developed countries is to provide government financial support for aeroengine development. There are adequate examples from the aeroengine manufacturing countries wherein the aeroengines have been designed and developed in the private sector (GE and P&W, USA) with the government's financial support, or the private companies have been taken over by the government (Gnome & Rhone, France, and formed into Snecma), which provided funds for the aeroengine development. In the United Kingdom, a private company facing a financial crisis was taken over by the government (Rolls-Royce, UK) for an infusion of government funds for aeroengine development, and, subsequently, after its successful development, the company was privatised again. It is ironic that in India, under the current political environment, the Indian government cannot directly fund a private company for research and development pertaining to any military project, such as the design and development of an aeroengine; however, if the project in its true sense is for the development of a low bypass turbofan core aeroengine, which can be equally beneficial for civil and military aeroengine development with the same core aeroengine, then probably the government would like to infuse the funds for its development. The aeroengine core will remain the same and can be used for civil as well as military aircraft with the addition of an afterburner and other minor modifications.

The complete sell-off of GTRE will enable the Indian company to harness the available talent and resources at GTRE, and have full authority and control mechanisms, which are considered necessary for the timely completion of the project. The capital infusion from the government will ensure its intent as a serious stakeholder for rapid development, testing, certification, and procurement for its utilisation of civil and military aircraft. The specification for the low bypass turbofan aeroengine may be prepared to keep the AMCA and the single-aisle civil and commercial aircraft of the Airbus 319/320/321 or Boeing 737 class for the transportation of 125-175 passengers in focus. Successful indigenous aeroengine development can be a real boon for the Indian government's aspiration for indigenous development of regional jet aircraft in the future. The specification should be generic, broad-based on currently validated technologies, achievable, and in accordance with contemporary aeroengines as of date. The specifications should not be altered until the basic aeroengine has completed its fabrication, integration, testing, validation, and certification processes. The government also needs to provide the necessary support from public sector companies such as HAL, BEL, BHEL, Heavy Engineering Corporation, Ranchi, Nuclear Fuel Complex, Hyderabad, Indira Gandhi Centre for Atomic Research (IGCAR), Kalpakkam, and various DRDO and CSIR laboratories. The government can also provide academic support from Indian Institutes of Technology (IITs) for fundamental research, algorithm validation, and an independent reliability engineering study of the aeroengine and its various modules. The terms and conditions of the lead company acquiring GTRE must be set most strictly in terms of its current business surplus, sound financial conditions, technological innovations, patent filling, and time-line of five years, including ground test, airborne test, and certification.

The government may also consider sanctioning a separate project for the indigenous design and development of an airborne platform for aeroengine testing that is not linked with the indigenous aeroengine development project. The development of aeroengines and airborne test platforms should continue in parallel so that the lead time for altitude tests and performance tests in the air within the country can be reduced considerably. The airborne test-bed platform should be a four-engine aircraft such as a Boeing 747 or Il-76 already in service in India with a trained crew for better reliability and adequately spacious to accommodate a number of gauges and display systems for observing and recording the critical aeroengine parameters for subsequent analysis. The task of airborne test-bed development can be assigned preferably to HAL since it has adequate experience in the installation and commissioning of a number of ground-based aeroengine test-beds indigenously and has inherent strength in aircraft structure modifications. The development of an airborne engine test-bed platform can also be done by private sector companies in a time-frame of two to three years; however, the company may require HAL's assistance for airframe modifications for the installation and integration of the aeroengine under test on the airborne platform aircraft. It may be difficult for the government to go through the fast-track divestment of GTRE since it is purely a research organisation with no return on investment until the successful completion of the project. Thus, Indian private sector companies may not be adequately motivated to acquire GTRE unless the government makes it a lucrative and viable business opportunity in terms of subsequent placement of purchase orders for future aeroengine procurements. The complete divestment process of GTRE may take a much longer time-frame, as we have witnessed with the divestment process of Air India in the recent past. Hence, this option is also not considered the best option for fast-track aeroengine design and development in India.

Option-4

Aeroengine Development by Public-Private Partnership on Collaborative Approach Through Risk and Revenue Sharing Basis
As we are aware, a number of Indian public and private sector entities have developed strong engineering strengths in the areas of design,

development, fabrication, and integration of complex technological machines with their innovative approaches in the various scientific segments such as materials, composites, aerospace structures, machining, process engineering, instrumentation, computer algorithms, software, etc. In the collaborative approach model, it is recommended that the aeroengine modules or sub-systems be developed through public sector undertakings and private sector companies based on their strong domain-dominant engineering knowledge and technical knowhow parallelly in order to shrink the overall developmental time-frame based on the basic specifications. A suitable selection and assessment criteria guideline may be compiled based on the brainstorming session with the representatives of the possible collaborators' companies in the public and private sectors. It will be a worthwhile idea to consider two companies for the development of critical modules and aggregates. A possible suggestive model is demonstrated below:

An aeroengine can be divided into the following major modules/sub-systems with the various sub-parts assembled inside it.

- Fan compressor (disc and blades).
- LP compressor (stator blades and disc with rotor blades).
- HP compressor (stator blades and disc with rotor blades).
- Combustion chamber (casing, fuel ring, burners, Ignitors, and flame stabilisers).
- HP turbine (nozzle guide vanes, disc, and blades).
- LP turbine (nozzle guide vanes, disc, and blades).
- Jet pipe and afterburner assembly (for fighter/combat aircraft).
- Jet nozzle.
- Main and auxiliary gearbox.
- FADEC.
- Aeroengine casing and bypass duct.
- Thrust reversal mechanism (for civil/commercial engines).

Some of the major components or aggregates installed on the aeroengine for its own functioning and also for providing mechanical, electrical, and hydraulic power sources to the aircraft need to be designed, developed, and prototype tested on their respective rig or test benches prior to their certification, validation, and integration into the main sub-systems. These are enumerated below:

- Gearbox with rotational speed reduction mechanism, comprising a set of gears to provide specific speed to run various services.
- Main fuel pump, metering unit and control system.
- Afterburner fuel pump (for combat or fighter aircraft).
- Thrust reversal mechanism for civil and commercial engines.
- The oil pump and scavenge pump for engine bearing lubrication and cooling.
- Air-oil radiator (heat exchanger for oil cooling).
- Main hydraulic pump.
- Auxiliary hydraulic pump.
- DC generator.
- AC generator.

The probable public sector and private sector companies that can be assigned the task of designing and developing the specific aeroengine module or sub-system, its test bench, and reliability study through an accelerated mission test based on their engineering strengths and rough assessments are suggested in the following table:

Sl. No.	Module/Sub-Sys/ Component	Public Sector Companies	Private Sector Companies
1.	Fan Module		
	Fan Blades	NAL, BHEL	TASL, L&T
	Fan Disc	DMRL/HEC, Ranchi MIDHANI	Bharat Forge, Kalyani Forge
2.	LP Compressor		
	LP Comp Disc	HEC, Ranchi MIDHANI	Bharat Forge, PTC Industries
	LP Comp Stator Blades	HAL (KPT) BHEL	Triveni Eng, Bengaluru
	LP Comp Rotor Blades	HAL (KPT) BHEL	Triveni Eng, Bengaluru

3.	HP Compressor		
	HP Comp Disc	HEC, Ranchi MIDHANI	Bharat Forge, PTC Industries
	HP Stator Blades	HAL (KPT) BHEL	Triveni Eng, TASL
	HP Rotor Blades	HAL (KPT) BHEL	Triveni Eng, TASL
4.	Combustion Chamber	MIDHANI ASP, Durgapur	Bharat Forge, L&T
5.	HP Turbine		
	HP Turbine Disc	MIDHANI HEC, Ranchi	Bharat Forge, PTC Industries
	HP Nozzle Guide Vanes	HAL (KPT) MIDHANI	Triveni Eng, Bengaluru
	HP Rotor Blades	HAL (KPT) MIDHANI	Triveni Eng, Bengaluru
6.	LP Turbine		
	LP Turbine Disc	MIDHANI HEC, Ranchi	Bharat Forge, PTC Industries
	LP Nozzle Guide Vanes	HAL (KPT), BHEL	Triveni Eng, Kun Aerospace
	LP Turbine Blades	HAL (KPT), BHEL	Triveni Eng, Kun Aerospace
7.	Jet Pipe and Afterburner Assembly	HAL (BC) HAL (KPT)	L&T
8.	Jet Nozzle	HAL (BC) HAL (KPT)	TASL, L&T
9.	Main and Auxiliary Gearbox		Gear & Gear Drives (India) Pvt Ltd, Bengaluru Anand Udyog, Belgaum
10.	FADEC		TCS, INFOSYS
11.	Aeroengine Casing and Bypass Duct	BHEL	Godrej Aerospace, Kun Aerospace, Chennai
12.	Main Fuel Pump, Metering Unit & Control Unit	HAL (LKO) CVRDE, Chennai	Tata Motors, Maruti-Suzuki Mahindra & Mahindra

13.	Afterburner Fuel Pump	HAL (LKO) CVRDE, Chennai	Tata Motors, Maruti-Suzuki Mahindra & Mahindra
14.	Oil Pressure Pump & Scavenge Pump	CVRDE, Chennai	Shanti Gear, Tata Motor, TVS
15.	Air-Oil Heat Exchanger	BHEL (Bharat Plate & Vessels)	Apollo Heat Exchanger, Mumbai
16.	Main Hydraulic Pump	BHEL	L&T, Dynaspede Enabletech
17.	Auxiliary Hydraulic Pump	BHEL	L&T
18.	DC Generator	BHEL	Siemens India, GE Power
19.	AC Generator	BHEL	Siemens India, GE Power
20.	Electrical Looms	HAL (LKO)	MKU
21.	HP & LP Rotors	HEC, Ranchi ASP, Durgapur	Bharat Forge, PTC Industries

The low bypass turbofan core aeroengine of 110-130 KN is to be designed by the lead private sector company, with capital assistance from the government, and the integration of the modules, sub-assemblies, and components will also be undertaken by the lead aeroengine integrator private company. However, the various modules, sub-assemblies, and components' functional and dimensional detail specifications will be issued centrally from the lead Indian integrator private company, and the module, sub-assembly, or component will be designed and developed by various public or private companies based on their merits, such as domain knowledge, expertise, machinery available, and reasonability of cost. Each of the modules, sub-assemblies, and components is to be designed, developed, and prototyped by the respective company selected for the same, and the necessary tools and test bench for individual modules, sub-assemblies, and component testing and self-certification must be completed strictly in accordance with the specified schedule. The validation test must be carried out by the specific quality manager of the respective company. The accelerated mission test, endurance test, and critical design review are also to be undertaken internally by the specified company. This aspect will reduce the total time span of the developmental process since the parallel development of all the modules, sub-assemblies, and components will progress simultaneously; the infrastructural cost will also reduce drastically, and the entire aeroengine developmental cycle can be accelerated. It is pertinent to note that the designated agencies will undertake the detailed design and generate manufacturing drawings and operational function simulations prior to actual fabrication so as to ensure the reduction of wastage in terms of material, time, and cost of fabrication. The fabrication will be undertaken by the designated company on the available machines, and if part outsourcing is required, the same can be done with proper documentation and traceability mechanisms. The following academic institutions and the research laboratory need to be associated with the project for expert professional consultations:

- Central Mechanical Engineering Research Institute, Durgapur.
- Central Vehicle Research and Development Establishment, Avadi.
- International Advance Research Centre for Powder Technology and New Materials, Hyderabad.
- Defence Metallurgical Research Laboratory, Hyderabad.
- Central Scientific Instruments Organisation, Chandigarh.
- Indira Gandhi Centre for Atomic Research, Kalpakkam.
- Indian Institute of Chemical Engineering, Hyderabad.
- National Metallurgical Laboratory, Jamshedpur.
- Nuclear Fuel Complex, Hyderabad.
- Indian Institute of Technology, Mumbai, Madras, Delhi, Kanpur, Varanasi and Kharagpur.
- Centre of Military Aircraft Airworthiness and Certification, Bengaluru.
- Directorate General of Civil Aviation, New Delhi.

The development of aeroengines is to be taken up as a project of national importance wherein the necessary assistance from all government and private enterprises is to be provided to the lead private company by all the agencies, and the necessary instructions may be issued by the Government of India in this regard to all the stakeholders. The project is to be considered a special project of national importance wherein all the participants should have adequate communication, frequent reviews, and ample discussion on the technological problem solving, and a single point of contact may be nominated by the government as well as the lead private sector integrator company for routine monitoring of the project progress and coordination with various agencies. Government funding may also be linked to the actual progress of the project on the ground, and an incentive of up to 20 per cent of the cost of the sub-project may also be provided for early completion of the project.

GTRE, Bengaluru, is also to be mandatorily engaged for its domain knowledge and also in the consultative and problem-shooting role, undertaking various reliability studies and ensuring availability of the test machineries. The lead private sector company will recommend the most suitable company in the public sector and private sector for the outsourcing task. The IAF should also remain actively associated with the project as the end user and part of the overall project monitoring group, along with the representative of the Civil Aviation Ministry, since the development of the aeroengine core will be a key factor for military as well as civil aircraft applications. The government may like to set up a monitoring committee also for the aeroengine testing platform in the air for altitude testing and other parameter validations so as to ensure that the air test-bed is freely available before the indigenous aeroengine is developed and getting ready for altitude testing. This is one of the most suitable proposals for the design, development, and prototype manufacturing of indigenous aeroengines; however, full-fledged support from the government and private sector enterprises must be provided by all the stakeholders, and the government's control, monitoring, and facilitation should be swift in case of any hold-up at any stage of development. This option is considered viable; however, continuous monitoring of the project's progress and facilitation by government agencies are to be ensured.

Option-5

Aeroengine Development in India in Mission Mode under the PMO
As covered earlier, it is of utmost importance that India opts, for the indigenous low bypass turbofan aeroengine development immediately in order to be recognised as an independent technology powerhouse internationally. It is pertinent to note that even 75 years after independence, India is completely dependent on foreign-based OEMs for the supply of aeroengines, spare parts, and even consumables for the Indian armed forces as well as the Indian civil aviation sector, which is a very vulnerable situation in the true strategic sense. Although India commenced its military aeroengine (Kaveri) development through GTRE, the aeroengine could not be successfully developed over the last three decades due to various reasons enumerated in the previous chapters, and the project was closed in 2014. It is understood that India has already spent over ₹2,100 crore or €240 million for the development of the Kaveri aeroengine without any success, and the project was closed in 2014. India could not derive any direct or technology offset benefit from the Kaveri engine project in terms of a new aeroengine or the reliability enhancement of vintage aeroengines powering the Indian Air Force aircraft. It is learnt that India is trying to develop or revive the Kaveri aeroengine development programme in consultation with Safran, the French company engaged in manufacturing world-class aeroengines for military and civil and commercial aircraft; however, the negotiations are failing over the technology and development costs aspects. As per an article in *The Economics Times* dated August 15, 2019, Safran agreed to adjust €250 million as part of an offset commitment, and the remaining €500 million was to be provided by DRDO. However, the price was not considered reasonable by DRDO, and the project did not get finalised. It is also learnt that Rolls-Royce, UK, and GE, USA, are also engaged in detailed discussions for the aeroengine development at GTRE; however, the same has not yet been confirmed so far. It is important to note that the development of an aeroengine, either with active support or as a joint venture or with ToT,

is possible much faster; however, it will impose a large number of severe restrictions that may or may not be acceptable to India as a sovereign state, as enumerated at Option 1 of this chapter.

Hence, under the present circumstances, it is considered most prudent that India re-start its full-fledged effort to develop the low bypass turbofan dual-use core aeroengine project under the overall umbrella of the PMO in line with ISRO. The entire aeroengine development programme under the PMO should preferably be undertaken through a public-private partnership to harness the full potential of the private sector's leadership, managerial skill, a higher degree of commitment, resolute determination, infrastructure, and resources, as well as the public sector's domain knowledge and rich engineering experience. The development cycle is estimated to take five years. The entire development cost must be borne by the government, in line with most of the advanced industrial countries that have successfully developed the aeroengine, since the aeroengine is a complex machine and the benefits derived from the country will be exponential. The government will recover the initial developmental cost and reap the benefits manifold in the form of taxes in the future. It is worthwhile to note that India has already placed orders for 40 LCA Mk-I and 83 LCA Mk-IA and is likely to procure at least 100 AMCA, for which development is in the advanced stage at ADA and HAL, Bengaluru.

Thus, India may spend more than US$ 20 billion on the purchase of an aeroengine for the LCA and AMCA during the life cycle of these aircraft. India has already signed a contract for 56 C-295 aircraft with Airbus Defence and Space at a cost of ₹21,935 crore as part of the ageing transport aircraft Avro-748. In all probability, the aircraft will be powered by a P&W, Canada, PW100. The PW 100 aeroengine has an SHP of 1800-5000. The proposed core aeroengine, with certain modifications, can also power the C-295 class of aircraft and other futuristic regional jets likely to fly from the upcoming UDAN project airports at tier-2 and tier-3 cities in India, where a large number of regional aircraft would be procured by Indian airline operators. It is contemplated that even the AN-32 aircraft, the workhorse of the IAF, may get replaced with the C-295 once it is indigenously produced. Thus, the same aeroengine (proposed indigenous aeroengine) may find further applications. Since it is proposed to develop a low bypass turbofan aeroengine with 110-130 KN thrust for military and civil aviation usages, it can be highly versatile, reliable, fully compliant with environmental emission laws, and fuel efficient in order to have a reasonable and affordable operating cost. The structure of the public-private sector company may be in the form of a joint venture or public-private partnership, or consortium. It may be located preferably in the Uttar Pradesh or Tamil Nadu defence corridor since a large number of defence-related industrial enterprises are likely to be located in these corridors, and two established industrial cities, Kanpur and Coimbatore, are already present in these corridors, respectively.

A proposed draft structure for the indigenous aeroengine development is given below:

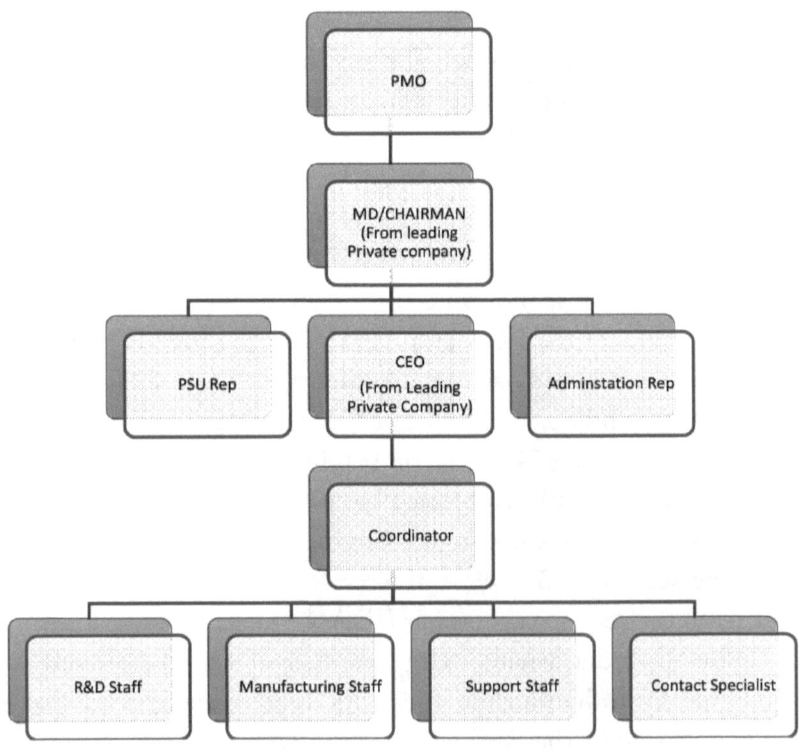

The organisation should be lean and mean, and the total staff should be the barest minimum since the entire task is to be performed preferably on an outsourcing model based strictly on capability assessment and domain knowledge of the industrial enterprise, independent of whether the firm is in the public or private sector. The specification of the aeroengine should be given by the IAF, and ADA combined teams in consultation with the Ministry of Civil Aviation. The Indian industry must be consulted prior to its finalisation of specifications. The private lead company shall design the aeroengine, and after validation of the design through simulation, the entire task of aeroengine development should be split into 20-25 modules, sub-assembly, or major components as described in Option-4 above in detail, and each module, sub-assembly, or major component should be outsourced to a different public or private sector company as per their engineering design and fabrication strengths and the domain knowledge of the respective company. The emphasis is on parallel design and development of each and every module, sub-assembly, and component, not sequentially, so as to reduce the developmental cycle time-frame and the cost factor. The quality aspect of each module, sub-assembly, or component will be accepted as per self-certification; however, the CEMILAC and DGCA representatives may be involved in all stages of aeroengine development in an advisory capacity only. The CEMILAC and DGCA should not be associated with vendor assessment, vendor selection, vendor communication, or internal company process inspection. In respect of all the aeroengine modules, sub-assemblies, and parts, the participating companies must be in possession of IS-9000, AS-9000C/D certification, and have six sigma trained professionals directly monitoring the quality aspects. The developmental process and progress monitoring system must be established, and continuous digital monitoring system must be adhered to. Any hold up or work restriction due to any reason, predictable or unpredictable, must be reported to the concerned officials, and the same must be resolved in the specified time-frame. The designs staff must constantly be in communication with the fabrication, manufacturing, and assembly staff for quick resolution of technical problems. The R&D staff may be selected solely on merit and calibre.

All the concerned agencies must make use of video conferencing and ample verbal and visual communication on secured modes, and physical meetings should be reduced to the barest minimum. Any bias, gender, age, reservation, serving, or retirement should not be the defined criteria for the selection of any scientific and engineering staff; however, a restrictive clause for premature exit must be applied in the interest of project secrecy, continuation, and completion on schedule.

Contract Specialists

It has been experienced in the past that most of a contract's documents pertaining to military as well as commercial airline aviation related product purchases are prepared by the foreign partners, and the draft copy is handed over to the Indian side for vetting and signature. The procurement staff, including the financial staff and the technical personnel, go through the main contract provisions and often clear the documents quickly in order to go for the contract faster. Since these personnel are not experts in contract management and legal provisions, they are ill-equipped to go through the fine print and understand the nuances of the contract provisions. The Indian side must train and induct contract managers with legal backgrounds to manage all future contracts and ensure unambiguous documentation to deal with post-contract management conflicts. This aspect is extremely important since the foreign vendors are well aware that the buyer is neither an expert on the new product's functionality nor on its mandatory spare requirements or guarantee conditionalities. They often draft the contract in their own favour and deny the legitimate right to the purchaser whenever a conflict of interest arises.

GTRE staff directly associated with the Kaveri aeroengine project may also be engaged in the related area of their specialisation. Specialists from the top academic institutions need to be engaged in fundamental research in dark areas such as composites development, special material superalloys, improvement of turbine blade resonance frequency, development of aeroengine control laws, FADEC, etc. Where the knowledge base is not available in India or the infrastructure, facility, machines, and test equipment are not available in India, the same need to be developed indigenously if the time-frame of development is longer. The indigenous solution may delay the entire project. At the same time, the indigenous effort must continue, and for urgent requirements, a limited quantity of specific equipment or the finished product may be bought outright from companies from abroad so as to ensure the timely completion of the project at all costs. The CEO may like to recruit consultants from India and abroad to assist the Indian designers' and manufacturing engineers' skill sets in order to resolve the various technical difficulties at the fastest pace.

The development of aeroengines under the PMO through a private company in a lead role and with government funds for R&D is considered the most effective option in order to fast track the developmental process by harnessing the best talent and infrastructure from the public and private sectors and academicians from the best institutions in India. The successful indigenous development of the aeroengine will boost the country's image in the international arena. It will ensure huge savings in foreign exchange resulting from the non-import of aeroengines for military and civil aviation segments. Further, the aeroengine can also be used for marine applications and power plants in the remote mountainous terrain in the Northeast region and the Jammu and Kashmir (J&K) areas, among other industrial applications. The project will also generate large-scale employment in the public as well as private sectors once the aeroengines enter the manufacturing phase post-certification, and India can even resort to the export of aeroengines to friendly foreign countries and earn huge foreign exchange in the future. The spinoff technologies will also benefit a large number of Small and Medium Enterprise (SME) companies, and they will attract a number of foreign OEMs to outsource most of their aeroengine component manufacturing to India, taking benefit of the low manufacturing costs in India in line with the auto industry's component sourcing from India. This option is considered the most desirable option for the time-bound indigenous aeroengine design, development, and prototype manufacturing project in India.

Conclusion

Indigenous design, development, and prototype manufacturing of low bypass 110-130 KN thrust modular core aeroengines are extremely important under *Aatmanirbhar Bharat* in order to reduce our dependence on foreign-based OEMs. Presently, all the aeroengines in the military and civil segments have been procured from OEMs abroad, and our dependence for spares, consumables, and lifetime technical support continues for the entire service life of the aeroengines (30-40 years). Most of these foreign-based OEMs have spent considerable resources in terms of money and time on the development of their aeroengines. Thus, they will not agree either for a joint venture or for co-development or joint development of the aeroengine in India. None of the OEMs will even agree to share the aeroengine design data since it will be tantamount to creating one more competitor in the international arena. The foreign-based OEMs, at best, can provide ToT for the manufacturing of the aeroengine in India; however, some of the most critical technologies will be retained by these OEMs and, thus, ensure our dependence for the entire service life of the aeroengine at considerable cost. Although India has made considerable progress in the arena of space research, information technology, road and rail transportation, quality steel production, automobiles, agriculture, manufacturing, communication, pharmaceuticals, and other industrial and financial segments, it is still completely dependent on foreign OEMs for aeroengines for the entire range of civil and military aircraft even after 75 years of independence. The Indian armed forces are operating a large number of combat aircraft from the erstwhile USSR, Russia, France, and the UK. Recently, India has also procured some transport aircraft and helicopters from the USA. In the arena of civil aviation, almost all the aircraft and aeroengines have been procured either from the USA or Europe. Thus, India is completely dependent and vulnerable for military air operations and civil transportation on foreign OEMs. India perforce must follow the alignment with these OEM countries; otherwise, there may be possibilities of sanctions being imposed on India, and the military, as well as the civil aviation segment, can be starved in terms of spares, consumables, and technical services, thereby adversely affecting the air power operations of military aircraft as well as the civil air transportation segment.

Although a project for the indigenous Kaveri aeroengine design and development was undertaken by GTRE under government funding to the tune of over ₹2,100 crore, the same could not be successfully completed even in more than three decades due to various reasons, as discussed in the previous chapters. It is imperative to fully utilise the resources created at GTRE and include the experienced and specialist personnel in the core team of the indigenous 110-130 KN turbofan modular core engine design and development. In view of the above, the requirement is to design and develop an indigenous aeroengine and air test-bed at the earliest possible time through the private sector, with government project funding, considering it as a project of national importance. The manufacturing of aeroengines in India will result in huge savings in foreign exchange to the tune of approximately US$ 25 billion in the next 15-20 years for the government and provide more than two million jobs directly and indirectly to Indian citizens in the span of 10-15 years.

Hence, it is considered imperative that a special project under the PMO for the design and development of indigenous low bypass 110-130 KN turbofan modular core aeroengines, which can be used by the military as well as the civil aviation segment with minor changes, be sanctioned with the highest priority. The proposed turbofan aeroengine is recommended to be designed and developed in the private sector for accelerated design and development through government funding in the form of an aeroengine developmental grant, which will make India a truly independent country and showcase its technological prowess to the world.

References

Rolls-Royce Plc, "The Jet Engines"

Pratt & Whitney, "The aircraft Gas Turbine Engine and its operation"

Irwin E. Treager, "Aircraft Gas Turbine Engine Technology"

MJ Cores & TW Wild, "Gas Turbine Engine"

Meherwan P. Boyce, "Gas Turbine Engineering Handbook"

Dr Wilfried Smarsly, "MTU Aero Engines"

RTO Technical Report, "More Intelligent Gas Turbine Engines"

Meinhard T. Schobeiri, "Gas Turbine Design, Components and System Design Integration"

M Nordqvist, Conceptual Design of a Turbofan

Malardalen University, Sweden, "Engine for a Supersonic Business Jet"

MRS Bulletin (October 2012), "Thermal Barrier Coatings for more efficient gas Turbine Engines"

Andry Y. Tkachenko, Venedikt S. Kuzmichev, Liia Krupenich, Victor Rybakov, "Gas Turbine Engine Optimization at Conceptual Designing"

R Kurz, K Brun, Cyrus Meher- Homji, Jeff Moore, F Gonzalez, "Gas Turbine Performance and Maintenance"

Rolls-Royce, "Gas Turbine in simple cycle & Combined Cycle applications"

Phillip P. Walsh, Paul Fleture, "Gas Turbine Performance"

Visser, WPJ and Broomhead, MJ, "Performance Prediction and Simulation of Gas Turbine Engine Operation"

Tulin Yildirim and (International Journal of Aerospace Engineering) Bulent Kurt, "Aircraft Gas Turbine Engine Health Monitoring System by Real Flight Data"

Na Zhang and Hai-Jun Xuan, "Investigation of high-speed rubbing behaviour of labyrinth honeycomb seal for turbine engine application"

Photographs

Fundamental of Jet Propulsion with Application Ronald D, Flack

Jet Propulsion, Third Addition, Nicholas Crumpsty and Andrews Heyes

Aircraft Power Plants, Ninth Edition, Thomas W.Wild

A History of jet propulsion, including Rockets, Raymond Friedman

Jet Engines , Klaus Hunecke

Design, Manufacturing and Operation of a small turbojet-Engine for research purposes, Ernesto & Stefano Giacometti

International Journal of turbo and Jet Engines, ISSN: 2191-0332

Integrated Jet Engine Engineering, SIEMENS

Modular Jet Engine Design: An alternate Power Generation Solution Lord Jonathan, Wam Bersie, Augustine

Aircraft Engine Design, Second Edition, Jack D. Mattingly, William H.Heiser David T. Pratt

www.gas-turbine-lab.mit.edu

www.degrayter.com

www.resourses.sw.siemens.com

www.cfmaeroengines.com

www.geaerospace.com

www.rolls-royce.com